Joint PhD degree in Environmental T

UNIVERSITÉ
— PARIS-EST

Docteur de l'Université Paris-Est

Spécialité: Science et Technique de l'Environnement

Dottore di Ricerca in Tecnologie Ambientali

Degree of Doctor in Environmental Technology

 Tampere University

Thesis for the degree of Doctor of Philosophy in Environmental Technology

PhD thesis –Väitöskirja – Proefschrift – Tesi di Dottorato – Thèse

Samayita Chakraborty

Biovalorisation of liquid and gaseous effluents of oil refinery and petrochemical industry

Defended on 12/12/2019, Paris

In front of the PhD evaluation committee

Prof. Rémy Gourdon	Reviewer
Prof. Mohammad Taherzadeh	Reviewer
Dr. Antonella Marone	Reviewer
Prof. Piet N.L. Lens	Promotor
Prof Christian Kennes	Co-promotor
Prof. Giovanni Esposito	Co-Promotor
Prof. Eric D. van Hullebusch	Co-Promotor
Prof Jukka Rintala	Co-Promotor
Prof. Mohammad Taherzadeh	Chair

Marie Skłodowska-Curie European Joint
Doctorate Advanced Biological Waste-to-
Energy Technologies
(ABWET)

Evaluation Committee

Chairperson

Prof. Mohammad Taherzadeh

Swedish centre for resource recovery

University of Borás

Sweden

Reviewers/Examiners

Prof. Rémy Gourdon

Department of Energy and Environmental Engineering

Institut National des Sciences Appliqueés de Lyon

France

Prof. Mohammad Taherzadeh

Swedish centre for resource recovery

University of Borás

Sweden

Dr. Antonella Marone

Italian National Agency for New Technologies, Energy and Sustainable Economic

Development

Italy

iii

Thesis promotor

Prof. Piet. N. L. Lens

Department of Environmental Engineering and Water Technology

UNESCO-IHE Institute for Water Education

The Netherlands

Thesis co-promotor

Prof. Giovanni Esposito

Department of Civil and Mechanical engineering

University of Cassino and Southern Lazio

Italy

Prof. Christian Kennes

Department of Chemical engineering and Bioengineering

University of A Coruña, Spain

Prof. Eric D. van Hullebusch

University of Paris-Est Marne-la-Vallée

France

Prof. Jukka Rintala

Faculty of Engineering and Natural Sciences

Tampere University, Finland

Supervisory committee

Thesis supervisor

Prof. Piet. N. L. Lens

Department of Environmental Engineering and Water Technology

UNESCO-IHE Institute for Water Education

The Netherlands

Thesis co-supervisor

Prof. Christian Kennes

Department of Chemical engineering and Bioengineering

University of A Coruña, Spain

Prof. Jukka Rintala

Faculty of Engineering and Natural Sciences

Tampere University, Finland

Thesis mentor

Dr. Eldon R. Rene

Department of Environmental Engineering and Water Technology

UNESCO-IHE Institute for Water Education

This research was conducted in the framework of the Marie Sklodowska-Curie European Joint Doctorate (EJD) in Advanced Biological Waste to Energy Technologies. This research was conducted under the auspices of the Graduate School for Socio-Economic and Natural Sciences of the Environment (SENSE). Partial research was also conducted by Xunta de Galicia, Spain for financial support to Competitive Reference Research Groups (GRC) (ED431C 2017/66) as well as Spanish ministry of economy, industry and competitiveness (MINECO) through project CTQ2017-8892-R.

CRC Press/Balkema is an imprint of the Taylor & Francis Group, an informa business

Published by:
CRC Press/Balkema
Schipholweg 107C, 2316 XC, Leiden, the Netherlands
Pub.NL@taylorandfrancis.com
www.crcpress.com – www.taylorandfrancis.com

ISBN: 978-0-367-61830-8

Contents

List of abbreviations

ATP   Adenosine triphosphate

AOR   Aldehyde oxidoreductase

AFOR   Aldehyde ferredoxin oxidoreductase

BDH   Butanol dehydrogenase

DMF   N, N-Dimethylformamide

DMA   N, N-dimethylacetamide

EPA   Environmental protection agency

FDH   Formate dehydrogenase

FT-IR   Fourier transform infrared spectroscopy

NADP   Nicotinamide adenosine triphosphate

NMR   Nuclear magnetic resonance

ppm   Parts per million

SEM   Scanning electron microscopy

TEM   Transmission electron microscopy

TBA   N-tert-butylacrylamide

UV   Ultraviolet

VFA   Volatile fatty acids

VOC   Volatile organic compounds

Acknowledgements

This work was supported by Marie-Sklodowska-Curie European Joint Doctorate (EJD) in Advanced Biological Waste to Energy (ABWET) funded by Horizon 2020 [grant number 643071], Xunta de Galicia, Spain for financial support to Competitive Reference Research Groups (GRC) (ED431C 2017/66) as well as Spanish ministry of economy, industry and competitiveness (MINECO) through project CTQ2017-8892-R.

This PhD spanning through 3 countries in 3 years , have only been possible with the continuous support, encouragement and positive criticism from my friends, colleagues, family and obviously my supervisors, co-supervisors and mentors. I would like to thank my PhD supervisor Prof. Piet N.L. Lens for his critical suggestions and insights. Dr. Eldon. R. Rene have been an optimistic and terrific mentor at every pitfall. Prof. Christian Kennes, my thesis co-supervisor, is like a father figure and true inspiration. Prof. Giovani, Prof, Eric and Prof Eric have been very crucial in suggesting the scientific aspects during summer schools. My fantastic friends from Netherlands, Tejaswini, Suniti, Iosif, Chris, Viviana, Joyabrata, Gabriele, Shrutika, Joyabrata, Angelo, Mirko, Lea, George, Joseph, Raghu, Ramita, Pritha and all the ABWET and ETECOS3 students, have always been by my side and I am ever grateful to have such wonderful time with them. Dealing with the language problem and getting adapted to new laboratory setup was only possible because of the lovely, helpful people, Pau, Kubra, Ruth, Borha and Anxela. My sincere thanks to the staff of IHE and TUT.

I would like to thank my Indian coworkers Dr. Ayan Dey, Dr. Arup Mandal, Dr. Rabin Bera for their continuous impetus to solve the research bottlenecks. Last, but not the least, my parents, especially my father and my family for their patience, criticism and unending support.

Summary

Liquid effluents of oil refinery contain toxic selenium oxyanions and phenol, while gaseous effluents contain toxic CO/syngas. To remove the phenol and simultaneously reduce the selenite oxyanions, a fungal-bacterial co-culture of *Phanerochaete chrysosporium* and *Delftia lacustris* was developed. Two modes of co-cultures of the fungus and the bacterium were developed. The first being a freely growing bacterium and fungus (suspended growth co-culture), the second being the growth of the bacterial biomass encircling the fungal biomass (attached growth co-culture). Both types of fungal-bacterial co-cultures were incubated with varying concentrations of phenols with a fixed selenite concentration (10 mg/L). The suspended growth co-culture could degrade up to 800 mg/L of phenol and simultaneously reduce 10 mg/L of selenite with production of nano Se(0) having a minimum diameter of 3.58 nanometer. The attached growth co-culture could completely degrade 50 mg/L of phenol and simultaneously reduce selenite to nano Se(0) having a minimum diameter of 58.5 nm.

In order to valorize the CO/syngas by bioconversion techniques an anaerobic methanogenic sludge was acclimatized to use CO as sole carbon substrate for a period of 46 days in a continuous stirred stank reactor, supplied with CO at 10 ml/min. 6.18 g/L acetic acid, 1.18 g/L butyric acid, and 0.423 g/L hexanoic acid were the highest concentrations of metabolites produced. Later, acids were metabolized at lower pH, producing alcohols at concentrations of 11.1 g/L ethanol, 1.8 g/L butanol and 1.46 g/L hexanol, confirming the successful enrichment strategy. The next experiment focused on the absence of trace element tungsten, and consecutively selenium on the previously CO acclimatized sludge under the same operating conditions. An in-situ synthesized co-polymeric gel of N-ter-butyl-acrylamide and acrylic acid was used to recover ethanol, propanol and butanol from a synthetic fermentation broth. The scope of repeated use of the gel for the alcohol recovery was investigated and every time approximately 98% alcohol was recovered.

Yhtenveto

Öljynjalostamon nestemäiset jätevedet sisältävät myrkyllisiä seleenioksianioneja ja fenolia, kun taas kaasumaiset jätevesit sisältävät myrkyllisiä CO/syngas. Fenolin poistamiseksi ja seleniittioksyanionien samanaikaiseksi pelkistämiseksi kehitettiin *Phanerochaete chrysosporiumin* ja *Delftia lacustris* -bakteerin sieni-bakteeri-yhteisviljelmä. Sienen ja bakteerin yhteisviljelmien kahta muotoa kehitettiin. Ensimmäinen on vapaasti kasvavia bakteereja ja sieniä (suspendoituneen kasvun yhteisviljelmä), toinen bakteerien biomassan kasvu, joka ympäröi sieni biomassan (kiinnittynyt kasvukulttuuri). Molempia tyyppisiä sieni-bakteeri-yhteisviljelmiä inkuboitiin vaihtelevien fenolipitoisuuksien kanssa kiinteällä seleniittikonsentraatiolla (10 mg / l). Suspendoitu kasvukulttuuri voisi hajottaa jopa 800 mg / l fenolia ja samanaikaisesti vähentää 10 mg / l seleniittia tuottamalla nano Se (0), jonka halkaisija on vähintään 3,58 nanometriä. Kiinnittynyt kasvatuskorppikotka voisi hajottaa kokonaan 50 mg / l fenolia ja seleniitin samanaikainen pelkistys nano Se: ksi (0), jonka vähimmäishalkaisija on 58,5 nm

CO / syngaskaasujen valorisoimiseksi biokonversiotekniikalla anaerobinen metaaniogeeninen liete aklimatoitiin käyttämään CO: ta ainoana hiilisubstraattina 46 päivän ajan jatkuvassa sekoitetussa säiliöreaktorissa, johon lisättiin CO nopeudella 10 ml / minuutti. 6,18 g / l etikkahappoa, 1,18 g / l voihappoa ja 0,423 g / l heksaanihappoa olivat korkein tuotettujen metaboliittien pitoisuus. Myöhemmin. hapot metaboloitiin alhaisemmassa pH: ssa tuottaen alkoholia konsentraatioissa 11,1 g / l etanolia, 1,8 g / l butanolia (41. päivä) ja 1,46 g / l heksanolia , mikä vahvistaa onnistuneen rikastusstrategian. Seuraava kokeilu keskittyi hivenainevolframin ja peräkkäisen seleenin puuttumiseen aikaisemmin CO-mukautetulle lietteelle samoissa käyttöolosuhteissa. . In-situ-syntetisoitua N-ter-butyyliakryyliamidin ja akryylihapon kopolymeerigeeliä käytettiin etanolin, propanolin ja butanolin talteenottamiseksi synteettisestä käymisliemestä. Geelin toistuvan käytön laajuutta alkoholin talteenottoa varten tutkittiin ja joka kerta noin 98-prosenttisen alkoholin voitiin todeta palautuvan.

Samenvatting

Het vloeibare effluent van olieraffinaderijen bevat de giftige stoffen selenium en fenol, terwijl het gasvormige effluent giftig CO/syngas bevat. Om fenol te verwijderen en tegelijkertijd de seleniet oxyanion concentratie te verminderen, werd een schimmel/bacterie co-cultuur van *Phanerochaete Chrysosporium* en *Delftia lacustris* ontwikkeld. Twee vormen van co-cultuur werden bestudeerd. De eerste bestond uit vrijgroeiende bacteriën en schimmels (co-cultuur van gesuspendeerde groei), de tweede uit groei van bacteriële biomassa die de schimmelbiomassa omgeeft (cultuur met gehechte groei). Beide typen co-cultuur van schimmel/bacterie werden geïncubeerd met variërende fenol concentraties bij een vaste seleniet (10 mg / L) concentratie. Een gesuspendeerde kweek kan tot 800 mg/l fenol afbreken en tegelijkertijd 10 mg/l seleniet reduceren door nano Se(0) met een diameter van ten minste 3,58 nanometer te produceren. De gehechte groei verwijderde gelijktijdig 50 mg/l fenol en seleniet, waarbij nano Se(0) met een minimale diameter van 58,5 nm werd gevormd.

Om CO/syngas te valoriseren door bioconversie werd een anaërobe methanogene slurry geacclimatiseerd om CO als het enige koolstofsubstraat te gebruiken gedurende 46 dagen in een continu geroerde tankreactor waaraan CO werd toegevoegd met een snelheid van 10 ml / minuut. De hoogste metabolietconcentraties waren 6,18 g/l azijnzuur, 1,18 g/l boterzuur en 0,423 g/l hexaanzuur. Later werden de zuren gemetaboliseerd bij lagere pH om alcohol te produceren in concentraties van 11,1 g/l ethanol, 1,8 g /l butanol en 1,46 g/l hexanol, hetgeen een succesvolle verrijkingsstrategie bevestigde. Het volgende experiment concentreerde zich op de afwezigheid van sporenelement wolfraam en selenium op een eerder met CO aangerijkte slurry onder dezelfde omstandigheden. Een in situ gesynthetiseerde N-tert-butylacrylamide en acrylzuurcopolymeergel werd gebruikt om het ethanol, propanol en butanol uit de synthetische fermentatievloeistof te winnen. De mate van herhaald gebruik van de gel voor alcoholterugwinning werd bestudeerd en elke keer werd ongeveer 98% alcohol teruggewonnen.

Sommario

Gli effluenti liquidi delle raffinerie di petrolio contengono ossianioni del selenio e fenoli tossici, mentre gli ffluenti gassosi contengono CO e syngas tossici. Al fine di eliminare i fenoli e allo stesso tempo ridurre gli ossianioni di selenio, è stata sviluppata una co-coltura batterica fungina di *Phanerochaete Chrysosporium* e *Delftia lacustris*. Sono state sviluppate due forme di co-colture di funghi e batteri. La prima è costituita da batteri e funghi a crescita libera (co-coltura sospesa), la seconda da una biomassa batterica adesa attorno alla biomassa fungina (co-coltura adesa). Entrambi i tipi di co-colture batteriche fungine sono stati incubati con diverse concentrazioni di fenolo e una concentrazione fissa di selenite (10 mg/L). La coltura sospesa è riuscita a degradare fino a 800 mg/L di fenolo, riducendo allo stesso tempo 10 mg/L di selenite con produzione di nanoparticelle di Se (0) di diametro pari almeno a 3,58 nanometri. La coltura adesa è riuscita a degradare completamente 50 mg/L di fenolo, riducendo contemporaneamente la selenite in nanopaticelle di Se (0) con un diametro minimo di 58,5 nm.

Per la bio-valorizzazione di CO / syngas una sospensione metanogenica anaerobica è stata acclimatata allo scopo di utilizzare la CO come unico substrato di carbonio per 46 giorni in un reattore a completa miscelazione additivato con 10 ml/minuto di CO . Le concentrazioni massime di metaboliti prodotti sono state 6,18 g/l di acido acetico, 1,18 g/l di acido butirrico e 0,403 g/l di acido esanoico. Successivamente, gli acidi sono stati metabolizzati a pH inferiore, producendo 11,1 g/l di etanolo, 1,8 g/l di butanolo e 1,46 g/l di esanolo, confermando il successo della strategia di arricchimento adottata. L'esperimento seguente si è concentrato sull'assenza di tungsteno in tracce e successivamente di selenio nel fango precedentemente acclimatato con CO nelle stesse condizioni operative. Un gel copolimerico di N-ter-butilacrilammide e acido acrilico sintetizzato in situ è stato usato per recuperare etanolo, propanolo e butanolo da un brodo di fermentazione sintetico. È stato approfondito l'uso ripetuto del gel per il recupero dell'alcol, ottenendo un'efficienza di recupero ogni volta pari a circa il 98%.

Résumé

Les eaux usées de raffineries de pétrole peuvent contenir des concentrations significatives et toxiques de sélénium sous forme d'oxyanions (séléniate ou sélénite) ainsi que du phénol, tandis que les emissions gazeuses de ces mêmes sites industriels peuvent contenir du monoxyde de carbone issu du syngas (Gaz de synthèse principalement composé d'hydrogène, monoxyde de carbone, méthane et dioxyde de carbone). Pour traiter le phénol et réduire simultanément les concentrations en sélénium, une co-culture mixte microchampignons-bactéries ou une co-culture simplifiée *Phanerochaete chrysosporium* et *Delftia lacustris* a été mis en oeuvre dans ce travail de thèse. Deux modes de croissance des co-cultures du microchampignon et de la bactérie ont été développés. Le premier est une co-culture bactérie(s)-microchampignon(s) en culture libre, le second est la croissance de la co-culture sur support. Les deux modes de co-cultures ont été incubés avec des concentrations variables ou constantes de phénol et du sélénium sous forme de sélénite à 10 mg / L. La co-culture libre a permis de dégrader jusqu'à 800 mg / L en phénol et traiter simultanément 10 mg / L de sélénite. Le sélénium étant principalement retrouvé sous la forme de sélénium élémentaire (Se(0)) nanoparticulaire ayant un diamètre minimum de 3,58 nanomètres. La co-culture fixée pouvant quant à elle dégrader complètement 50 mg / L de phénol tout en réduisant simultanément le sélénite en sélénium élémentaire (Se(0)) nanoparticulaire ayant un diamètre minimum de 58,5 nm.

Afin de valoriser le monoxyde de carbone du syngas par des techniques de bioconversion, des boues méthanogènes anaérobies ont été acclimatées pour utiliser le monoxyde de carbone comme unique substrat carboné pendant une période de 46 jours dans un réacteur sous agitation continue alimenté en monoxyde de carbone à 10 mL / min. Les plus fortes concentrations de métabolites produits étaient les suivantes: 6.18 g / L d'acide acétique, 1.18 g / L d'acide butyrique et 0,423 g / L d'acide hexanoïque. Plus tard, les acides ont été métabolisés à un pH inférieur, produisant des alcools à des concentrations de 11,1 g / L d'éthanol, 1,8 g / L de butanol et 1,46 g / L d'hexanol, confirmant ainsi le succès de la stratégie d'enrichissement. L'expérience suivante portait sur l'effet de la carence d'éléments traces comme le tungstène et le sélénium sur les boues acclimatées au CO dans les mêmes conditions de fonctionnement. Un gel de copolymères synthétisé in situ à partir du N-ter-butylacrylamide et de l'acide acrylique a été utilisé pour récupérer l'éthanol, le propanol et le butanol à partir d'un bouillon de fermentation synthétique. L'utilisation répétée du gel pour la récupération d'alcools a été étudiée et chaque fois environ 98% d'alcools formés ont été récupérés.

List of publications

1: S Chakraborty, ER Rene, PNL Lens, MC Veiga, C Kennes 2019. Enrichment of a solventogenic anaerobic sludge converting carbon monoxide and syngas into acids and alcohols. Bioresource technology 272, 130-136.

2. S Chakraborty, ER Rene, PNL Lens 2019. Reduction of selenite to elemental Se(0) with simultaneous degradation of phenol by co-cultures of *Phanerochaete chrysosporium* and *Delftia lacustris*. Journal of Microbiology 57 (9), 738-747.

3. S Chakraborty, R Bera, A Mandal, A Dey, D Chakrabarty, ER Rene, PNL Lens 2019. Adsorptive removal of alcohols from aqueous solutions by N-tertiary-butylacrylamide (NtBA) and acrylic acid co-polymer gel Materials Today Communications, 100653.

4. S. Chakraborty, ER Rene, PNL Lens, MC Veiga, C Kennes. Effect of tungsten and selenium for CO and syngas bioconversion by enriched anaerobic sludge. (In preparation).

Author's contributions

Paper 1: S. Chakraborty performed the experiments, interpreted the data and wrote the manuscript. C. Kennes helped in the conceptualization, execution of experiments, data analysis and revising manuscripts. M.C. Veiga helped in giving the laboratory facility and helped in solving experimental difficulties. E.R. Rene and P.N.L. lens helped in planning experiments, data analysis and revising manuscripts.

Paper 2: S. Chakraborty performed the experiments, interpreted the data and wrote the manuscript. E.R. Rene and P.N.L. Lens helped in planning experiments, data analysis and revising manuscripts.

Paper 3: S. Chakraborty performed the experiments, interpreted data and wrote the manuscripts. A. Dey, Arup Mandal, Rabin Bera contributed to the planning the experiments, revising manuscripts and correcting the proofs. E.R. Rene, P.N.L. Lens and D. Chakraborty helped in planning experiments, data analysis and revising manuscripts.

Paper 4: S. Chakraborty performed the experiments, interpreted the data and wrote the manuscript. C. Kennes helped in the conceptualization, execution of experiments, data analysis and revising manuscripts. M.C. Veiga helped in giving the laboratory facility, infrastructure and also helped in solving experimental difficulties. E.R. Rene and P.N.L. Lens helped in planning experiments, data analysis and revising manuscripts.

Chapter 1

General Introduction

1. Introduction

1.1 Background

The world's incessantly increasing demand for fossil fuels and technological advancements are driven by population growth and never ending rise in aspirations for improving people's lifestyle and standard of living. The modern amenities of life in combination with the inevitable need for transport fuels have accelerated the growth of the petrochemical industries, including oil refineries and their by-products. Other petroleum based products like varnishes and cosmetics have phenolic compounds. The application of petro-based products has escalated fast and an apprehension has developed regarding the complete depletion of its natural reserves (Lindholt et al., 2015). In addition, a callous attitude by the appropriate authorities towards the nuisance created by the large variety of hazardous wastes generated by the petro-based industries has endangered not only the human civilization, but also the marine and aquatic life forms, thereby hampering respiration and the growth of different aquatic species (Abdelwahab et al., 2009).

The liquid waste streams effusing from petrochemical industries mostly include phenolic compounds with other polyaromatic hydrocarbons and a considerable proportions of toxic ions like selenite (Werkeneh et al., 2017). The accidental release of these compounds to water bodies could pose a serious threat to human health and the environment. The gaseous emissions produced mostly by incineration of the petro-products include harmful compounds like syngas (a toxic mixture of CO, CO_2 and H_2 at varying compositions). For example, CO can cause dizziness, vomiting, unconsciousness and even death (Prockop et al., 2007). It is also a major gas produced from steel plants by the incomplete combustion of any carbonaceous feedstock and in bio-refineries (Molitor et al., 2016). Recent investigations are based only on the treatment of the oil refinery wastewater including physico-chemical processes like reverse osmosis, adsorption, ultrafiltration, chemical destabilization, membrane processes and also biological anaerobic treatment for hydrocarbon removal (Varjani et al., 2019). Petrochemical refineries can serve as a great 'waste to energy resource' as the gaseous emission contains syngas (CO, CO_2, and H_2) and liquid waste contains the hydrocarbons with various metalloids like selenium. The biological conversion of the liquid wastes and gaseous emissions of petrochemical refineries offer paramount scope to produce value added products (Psomopoulos et al., 2009). The biological approach ensures a comparatively lower energy intensive, eco-friendly way of detoxification of these harmful compounds. Using the well-known Wood-ljungdahl pathway (Fernández-Naveira et al., 2017), *Clostridium* sp. produces industrially relevant acids and alcohols which can be used as biofuels. Industrial waste is thus converted to biofuels which signifies the concept of biological waste to energy conversion.

1.2 Problem statement

Phenol is a toxic and carcinogenic aromatic hydrocarbon used in process industries, e.g. dye-manufacturing, coke-oven, fiber-glass manufacturing, pulp and paper industries, phenolic resin synthesis, plastics and varnish industries (Gangopadhyay et al., 2018). They are also formed as an intermediate in the pharmaceutical and herbicide industry. It is difficult to remove this important pollutant due to its high reactivity towards a wide range of compounds, ready convertibility to some other compounds which are sometimes isomeric in nature, high energy is involved in its removal and easy absorption by human skin and cell membranes. The aqueous phenolic effluents when contaminated with soluble toxic oxyanions of metalloids, mostly encountered in oil refinery effluents (Lawson and Marcy 1995) pose great problems to the life forms because of the synergism in toxicity of phenol imposed by intrinsically poisonous and toxic metalloids.

Most of the phenolic effluents are released into water bodies and phenol concentrations of $\sim$ 2.85 mg/L and 4.11 mg/L have been observed to drastically reduce the dissolved oxygen content, primary productivity, phytoplankton and zooplankton populations (Saha et al., 1999). Detailed research on the interaction between phenol and metalloids are yet to be sought into for a more transparent understanding of the combined negative effects of phenol and metalloid. However, their co-existence is a real threat to the environment. Substantial research activities have already been carried out to remove these toxicants, focusing on only phenol (Villegas et al., 2016), which include: (i) physico-chemical treatment processes such as distillation, adsorption, extraction, electrochemical treatment, (ii) membrane processes such as reverse osmosis, nanofiltration, pervaporation and membrane distillation, (iii) advanced oxidation processes such as UV/H_2O_2 treatment, Fenton and photo-Fenton based processes, wet air oxidation and ozone treatment, and (iv) biological treatment systems that include aerobic and anaerobic processes. Most of the processes applied to remove the toxicants are chemical processes which produce harmful toxic by-products, especially for phenol. Among these treatment technologies, biological treatment has been shown to be less energy intensive than chemical treatment systems. However, the complete removal of phenol from the wastewater still remains a big concern.

On the other hand, the removal of selenium oxyanions from wastewater is also influenced by several physico-chemical and biological conditions (Lenz et al., 2009). According to the literature, biological techniques are mainly based on removing either only phenol (Abdelwahab et al., 2009) or only selenite oxyanions in the presence of a favorable carbon substrate depending on the type of biocatalyst used (Espinosa-Ortiz et al., 2015). The removal is sometimes not complete and the technique is not cost-efficient. Thus, it is necessary to investigate the possibility of eco-friendly, cheap bioremediation techniques that can eliminate these two toxicants from the systems. The only dual detoxification of phenol and selenite was reported in an up-flow fungal bioreactor and the maximum phenol concentration tested in that study was only 400 mg/L (Werkeneh et al., 2017). An efficient microbial association, i.e. syntrophy between the fungus and the bacterium can build the foundation of the biological removal process (Deveau et al., 2018) which is the rationale behind the experiments undertaken. In aerobic environment, where the refinery wastewater is released, the successful fungal-bacterial association is a common phenomenon. This co-metabolic interaction has been exploited in the present work for the detoxification of phenol and selenite ions present in refinery wastewater. Moreover, different physical interactions between bacteria and fungus have been studied in the present dissertation that emphasizes the novelty of this research work.

CO is a major compound found in waste gases from steel industries and from biorefineries in the form of syngas (Abubackar et al., 2011). It is also produced from the gasification of biomass, solid waste or another carbonaceous feedstock. In the perspective of a fossil fuel crisis, a strong impetus is being felt for the development of energy efficient and cost-effective renewable sources of energy from waste gases. Methane is an alternative for gaseous biofuel (Tilche and Galatola 2008). But, it imparts a green-house effect which is 84 times than that of carbon dioxide. Complete utilisation of methane as biofuel in plants and broilers without faulty operations, seem quite difficult. Thus accidental leakage of methane may contribute to the global warning. The financial times quote "Scientists warn over record levels on methane" on May 24, 2019 ('Record methane levels" 2019). So, an alternative fuel is the need of the hour. Alcohols are potential biofuels which does not contribute to the global warming (Agarwal, 2007). The Fischer-Tropsch catalytic process of syngas conversion to alcohols, mainly ethanol is still the industrial process of alcohol production. Chemical processes like these are faster than biological approaches, but the advantages are (i) Near complete conversion efficiencies due to the irreversible nature of biological reactions (Klasson et al., 1991, 1992), (ii) the high enzymatic specificities of biological conversions also result in higher product selectivity with the formation of fewer by-products.(iii) Poisoning of biocatalysts by sulfur, chlorine and tars other than inorganic catalysts (Michael et al., 2011; Mohammadi et al., 2011), is less feasible which reduces the gas pre-treatment costs. In recent years, many bio techniques have been developed for the bioconversion of syngas/CO into efficient biofuels with reasonable calorific value and without any impurities that might otherwise affect the fuel quality (Diender et al., 2016, Abubackar et al., 2018, Fernández-Naveira et al., 2017).

Although different syngas/CO fermentation systems have shown good process efficiencies, the presence of syngas impurities, at varying concentrations, affects the cell growth, enzyme activity and various other parameters pertinent to the bioconversion techniques employed. These impurities include compounds such as hydrogen sulfide (H_2S), sulfur dioxide (SO_2), ammonia (NH_3), nitrogen (N_2), methane (CH_4), acetylene (C_2H_2), carbonyl sulfide (COS), oxygen (O_2), water (H_2O), chlorine compounds, mono-nitrogen oxides (NO_x), ethylene (C_2H_4), ethane (C_2H_6), benzene (C_6H_6), hydrogen cyanide (HCN), ash and tar (Xu et al., 2011) that are mainly produced depending on the syngas composition, source and the fermentation conditions used.

From a practical viewpoint, there are several challenges to be addressed, in order to fully utilize the energy content of syngas. However, the mixed culture based biological processes by virtue of their several advantages like inexpensive biocatalysts, ability to withstand fluctuating process conditions, higher tolerance to the syngas impurities and high product selectivity are assumed to possess high potentialities associated with syngas conversion. Mixed cultures comprising of anaerobic sludge are easily acclimatized and resilient to the toxic impurities. Few research has focused on mixed cultures due to the narrow range of microorganisms which are able to acclimatize to CO and capable of convert CO to medium chain carbon compounds like low molecular weight acids and alcohols. This research scenario is the rationale for enriching a mixed culture (anaerobic sludge) for CO fermentation, experimenting with supplementation of crucial medium components.

In continuation with this progressive research line on developing high performing CO/syngas fermentation systems, studying the influence of trace elements on the process efficiency is considered to be of paramount importance. Some of the metals like tungsten or metalloids like selenium and rare earth elements and are found to be involved in the formation of acids from syngas. However, the complete absence of these individual elements would likely to inhibit, partially or completely, the process of metabolism. In mixed culture systems, where different microbes have different enzymes and their co-factors contain different metals , experiments performed with the individual absence of some particular trace elements are likely to focus on important metabolic flux of CO fermentation producing acids and alcohols. This area of research is very interesting as some metals which are necessary cofactor for one microorganism, may inhibit or may have no effect at all on another microorganisms. Thus, selenium has no effect on alcohol production by CO fermenting *Clostridium autoethanogenum*, but addition of selenium enhances production of alcohols by *Clostridium ragsdaleii*. This bioconversion of syngas into biofuels produces a mixture of alcohols, mostly of low molecular weights ranging from ethanol to butanol and sometimes a mixture of hexanol and acids (Fernández-Naveira et al., 2017). Consecutive deficiency of tungsten and selenium in mixed culture CO fermentation would shed some light on the flux of CO fermentation and the array of metabolites produced.

Considering downstream processing and recovery of the end-products, it is a challenge to separate the different alcohols. The conventional methods for the recovery of alcohols from a fermentation broth include pervaporation (Qureshi et al., 1990), extraction (Jiang et al., 2009), gas stripping (Cai et al., 2016) and adsorption (Xue et al., 2016). Among these, adsorption is the most cost-effective and less energy intensive technique. Thus, the quest for a suitable adsorbent/absorbent for the selective separation of alcohols like ethanol, propanol, butanol and hexanol from the fermentation broth will offer more practical application of the produced biofuels.

1.3 Research objectives

Considering the scope of this work, the primary research objectives of this thesis can be stated as follows:

1. Evaluate the simultaneous removal of phenol and selenite ions from wastewater using microorganisms:

i) Determine the dual detoxification capacity of the fungus *Phanerchaete chrysosporium* and the bacterium *Delftia lacustris* by two different modes of growth: a) Suspended growth (in isolation) co-culture of the fungus and the bacterium. b) Attached growth co-culture of the fungus and the bacterium.

ii) A comparative study of the two growth systems, particularly in terms of its efficiencies with respect to the extent of phenol removal and selenite reduction along with the mode of biomass formation to assess the synergistic microbial metabolism involved.

2. Enrichment of anaerobic sludge for solventogenic bacteria which is able to ferment C1 gases, i.e., CO, CO_2 and syngas to ethanol and higher alcohols and the operating parameters are as follows.

i) The influence of pH variation in a continuous gas-fed reactor on the amount of metabolites produced including acids and alcohols.

ii) The effect of addition of a specific inhibitor of methanogens is in order to select and facilitate the growth of solventogens

iii) The effect of yeast extract and L-cysteine-HCl addition on the metabolism of solventogenic acetogens

3. Effect of trace metal (selenium Se or tungsten, W) addition on the CO conversion efficiency by CO adapted anaerobic sludge.

i) Effect of Se and W on production of metabolites.

ii) Gas consumption efficiency and effect of pH on the CO conversion.

4. An endeavor has been made to separate the biofuels (mostly ethanol and butanol) from the fermentation broth by the process of adsorption where a novel gel has been used as an adsorbent. A study has been undertaken to determine.

i) The adsorption-absorption and desorption process of the selected alcohol on the gel surface.

ii) The mechanism of alcohol imbibition into the gel core and retaining it for subsequent use.

iii) The mode of alcohol recovery from the gel and its reuse.

1.4 Structure of the PhD thesis

Figure 1.2 overviews the structure of this PhD thesis. Chapter 1 gives a brief introduction of the whole research work. There is a concise description about the burning problem of energy crisis and the applicability of the research undertaken here to address the different aspects. The specific research objectives with the structure of the thesis have been framed with a discussion of the issues addressed in every chapter. Chapter 2 gives a detailed research performed before on the relevant fields focused in the different chapters. A brief overview on degradation of phenol and other phenolic compounds with the presence of simultaneous metal ions, have been documented. The current research parameters affecting CO fermentation, have also been discussed.

Chapter 3 explained the proof of concept of a fungal bacterial co-culture for efficient detoxification of phenol and selenite ions. Two different modes of growth of the fungal-bacterial co-culture were investigated using different nutrient medium. Different concentrations of phenols were tested to find the adaptability and detoxification efficiency of the two co-cultures. As a by-product, production of nano Se(0) were also observed.

Chapter 4 describes the enrichment of methanogenic sludge to solventogenic sludge for production of alcohols. Operation parameters like pH were varied to stimulate acids and alcohol production. Simultaneously, the concentration of components of nutrient medium like yeast extract and L-cysteine-HCl were also modulated to find the production profile of acids and alcohols. Chapter 5, in continuation of chapter 4, explains how the individual absence of selenium and tungsten affects the production of acids and alcohol. CO uptake efficiency by the CO/syngas fermenting microbes was investigated. Denatured gradient gel electrophoresis also revealed some of the species involved in the fermentation process.

Chapter 6 elucidates the synthesis of a polymeric gel and its efficiency in recovering alcohols from model aqueous solutions. Ethanol, propanol and butanol were used at definite concentrations as model compounds to be removed and recovered by the polymeric gel. Cyclical swelling and de-swelling of the N-tertiary-Butyl-Acrylamide/Acrylic Acid gel in ethanol, propanol and 1-butanol for two consecutive cycles were observed.

Chapter 7 discusses the output of the research performed and possible future perspectives. Future perspectives of this work includes scaling up of the processes undertaken. Moreover, detailed microbiological analysis is also necessary for a better understanding of the process.

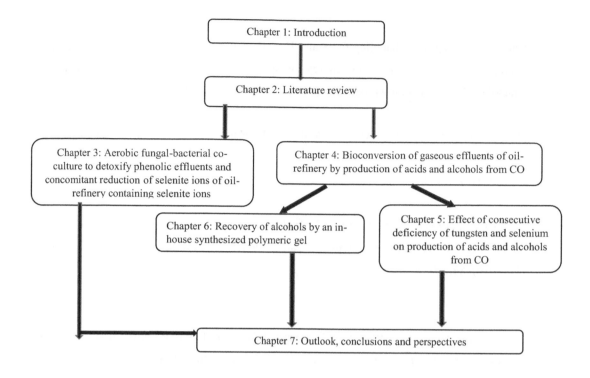

Figure 1.2 Overview of the chapters in this PhD thesis

References

Abdelwahab, O., Amin, N.K., El-Ashtoukhy, ES Z. Electrochemical removal of phenol from oil refinery wastewater. J. Hazard. Mater. 163 (2-3) (2009): 711-716.

Abubackar, H.N., Veiga, M.C., Kennes.C. Biological conversion of carbon monoxide: rich syngas or waste gases to bioethanol. Biofuels Bioprods. Bioref, 5 (2011): 93-114.

Abubackar, H. N., Veiga, M.C., Kennes, C. Production of acids and alcohols from syngas in a two-stage continuous fermentation process. Bioresour. Technol. 253 (2018): 227-234.

Agarwal, A.K. Biofuels (alcohols and biodiesel) applications as fuels for internal combustion engines. Prog. Energ. Combust. 33 (3) (2007): 233-271.

Cai, D., Huidong, C., Changjing C., Song, H., Yong, W., Zhen, C., Qi, M. Gas stripping–pervaporation hybrid process for energy-saving product recovery from acetone–butanol–ethanol (ABE) fermentation broth. Chem. Eng. J. 287 (2016): 1-10.

Deveau, A., Gregory, B., Jessie, U., Mathieu, P., Matthias, B., Saskia, B., Stéphane, H. Bacterial–fungal interactions: ecology, mechanisms and challenges. FEMS Microbial. Rev. 42 (3) (2018): 335-352.

Diender, M., Alfons, J.M.S., Sousa, D.Z. Production of medium-chain fatty acids and higher alcohols by a synthetic co-culture grown on carbon monoxide or syngas. Biotechnol. Biofuels 9 (1) (2016): 82.

Espinosa-Ortiz, E.J., Rene, R. E., van Hullebusch, E.D., Lens, P.N.L. Removal of selenite from wastewater in a *Phanerochaete chrysosporium* pellet based fungal bioreactor. Int. J. Bidet. Biodeg. 102 (2015): 361-369.

Fernández-Naveira, Á., Veiga, M.C., Kennes, C. H-B-E (hexanol-butanol-ethanol) fermentation for the production of higher alcohols from syngas/waste gas. J. Chem. Technol. Biotechnol. 92 (4) (2017): 712-731.

Gangopadhyay, U. K., Dongre, S. S., Salunkhe, P.R. Biological method to reduce phenol content for efficient and environment friendly effluent treatment. Man-Made Tex. in Ind. 46 (11) (2018): 367-371.

Jiang, Bo, Zhi-Gang, L., Jian-Ying, D., Dai-Jia, Z., Zhi-Long, X. Aqueous two-phase extraction of 2, 3-butanediol from fermentation broths using an ethanol/phosphate system. Process Biochem. 44 (1) (2009): 112-117.

Lawson, S., Macy, J.M. Bioremediation of selenite in oil refinery wastewater. Appl. Microbiol. Biotechnol. 43 (1995): 762-765.

Lenz, M., Lens, P.N.L. The essential toxin: the changing perception of selenium in environmental sciences. Sci. Tot. Env. 407 (12) (2009): 3620-3633.

Lindholt, L. The tug-of-war between resource depletion and technological change in the global oil industry 1981–2009. Energy 93 (2015): 1607-1616.

Klasson K. T., Ackerson M. D., Clausen E. C., Gaddy J. L. (1991). Bioreactor design for synthesis gas fermentations. Fuel 70 (1991): 605–614.

Klasson K. T., Ackerson M. D., Clausen E. C., Gaddy J. L. Bioconversion of synthesis gas into liquid or gaseous fuels. Enzyme Microb. Technol. 14 (1992): 602–608.

Michael K., Steffi N., Peter D. The past, present, and future of biofuels – biobutanol as promising alternative, in Biofuel Production-Recent Developments and Prospects, ed dos Santos Bernades M. A., editor. (Rijeka: Intech) 15 (2011): 451–486.

Mohammadi M., Najafpour G. D., Younesi H., Lahijani P., Uzir M. H., Mohamed A. R. (2011). Bioconversion of synthesis gas to second generation biofuels: a review. Renew. Sustain. Energy Rev. 15 (2011): 4255–4273.

Molitor, B., Richter, H., Martin, M.E., Jensen, R.O., Juminaga, A., Mihalcea, C., Angenent, L.T. Carbon recovery by fermentation of CO- rich off gases-turning steel mills into biorefineries. Bioresour. Technol. 215 (2016): 386-396.

Prockop, L.D. and Chichkova, R.I. Carbon monoxide intoxication: an updated review. J. Neurological Sci. 262 (1-2) (2007): 122-130.

Qureshi, N., Blaschek, H.P. Butanol recovery from model solution/fermentation broth by pervaporation: evaluation of membrane performance. Biomass. Bioenerg. 17 (20) (1999): 175-184.

Record methane levels pose new threat to Paris accord. (2019, May 24). Retrieved from URL "https://www.ft.com/content/9a3c0514-7d6b-11e9-81d2-f785092ab560"

Saha, N.C., Bhunia, F., Kaviraj, F. Toxicity of phenol to fish and aquatic ecosystems. Bull. Environ. Contam. Toxicol. 63 (2) (1999): 195-202.

Tilche, A., Galatola, M. The potential of bio-methane as bio-fuel/bio-energy for reducing greenhouse gas emissions: a qualitative assessment for Europe in a life cycle perspective. Water Sci. Technol. 57 (11) (2008): 1683-1692.

Varjani, S., Joshi, R., Srivastava, V.K., Ngo, H.H., Guo, W., Treatment of wastewater from petroleum industry: current practices and perspectives. Environ. Sci. Pollut. R. (In press) (2019): 1-9.

Villegas, L.G.C., Mashhadi, N., Chen, M., Mukherjee, D., Taylor, K. E., Biswas, N. A short review of techniques for phenol removal from wastewater. Curr. Pollut. Rep. 2(3) (2016): 157-167.

Werkeneh, A.A., Rene, E.R., Lens, P.N.L. Simultaneous removal of selenite and phenol from wastewater in an up flow fungal pellet bioreactor. J. Chem. Technol. Biotechnol. 93 (2017): 1003-1011.

Xu, D., Tree, D.R., Lewis, R.S. The effects of syngas impurities on syngas fermentation to liquid fuels. Biomass Bioeng. 35 (7) (2011): 2690-2696.

Xue, C., Liu, F., Xu, M., Tang, I.C., Zao, J., Bai, F., Yang, S.T. Butanol production in acetone-butanol-ethanol fermentation with in situ product recovery by adsorption. Bioresour. Technol. 219 (2016): 158-168.

Chapter 2

Literature review

2.1 Bioremediation of phenol and selenite in effluents of oil refineries and the

petrochemical industry

Phenol is a toxic (Kavitha and Palanievelu, 2004) and stable compound that bio accumulates in the environment. The Environmental Protection agency (EPA) has set the maximum permissible limit of phenol discharge in wastewater to be less than 1 ppm (1 mg/L) (Abdelwahab et al., 2009). Phenol is one of the volatile organic compounds (VOCs) present in almost every petrochemical plant's effluent. The concentration of phenolic compounds in the liquid effluents of oil refinery and petrochemical industry can be in the range of 20–200 ppm (Hernández-Francisco et al., 2017). The main sources of total phenols in the received waste streams at ENOC-RWTP (Emirates National oil company-refinery wastewater treatment plant) are the tank water drain (average 11.8 mg/l), the desalter effluent (average 1.4 mg/l), and the neutralized spent caustic (average 234 mg/l) waste streams (Al Hashemi et al., 2015). Generally, phenol is mainly produced by the process of hydrolytic cracking from oil refinery and petrochemical industry.

Several physical processes have been used to remove phenol from oil refinery effluents. Activated carbon has been used to remove phenol from petrochemical wastewater (El-Nas et al., 2010). Electrocoagulation has been applied to remove 97% of phenol after 2 hours at high current density of 23.6 mA/cm^2 and pH 7 at the highest concentration of 30 mg/L phenol (Abdelwahab et al., 2009). Due to costly and energy intensive physical processes, biological degradation of phenol is more favourable (Pradeep et al., 2015). Complete degradation of about 35 mg/L phenol in oil refinery effluent by a co-culture of *Pseudomonas aeruginosa* and *Pseudomonas fluorescen*ce has been reported (Ojumu et al., 2005). Recently, two plant species, *Typha domingensis* and *Leptochloa fusca*, in association with a consortium of crude oil-degrading bacterial species *Bacillus subtillis* LOR166, *Klebsiella* sp. LCRI87, *Acinetobacter Junii* TYRH47 and *Acinetobacter* sp. BRSI56 have been found to degrade 95% of hydrocarbons, including phenol (Rehman et al., 2019).

The biological route of phenol degradation with simultaneous metal reduction, particularly chromium [Cr(VI)], a metal present in oil refinery effluents, has been investigated quite extensively (Ontañon et al., 2017, Gupta et al., 2015). This dual detoxification process could also be applied to other metalloids like selenium oxyanions, which are also present in oil refinery effluents (Lawson and Marcy, 1995). However, so far there is only one report on the simultaneous degradation of phenol and the reduction of selenite ions (Werkeneh et al., 2017).

Selenium is an essential metalloid belonging to the chalcogen group 16 (VIA) in the periodic table which has a very narrow range of being an essential nutrient (50-70 µg/day) (Kipp et al., 2015) and a toxic element. The oxyanions (+VI and +IV) of Se are soluble and mobile, while the elemental form Se(0) is insoluble in water (Chasteen and Bentley 2003). Selenium oxyanions alter their speciation through the oil refinery and selenium removal is a difficult process. Figure 2.1 shows the fate of selenium from the start to the end in an oil refinery. The removal process includes an iron-selenium co-precipitation process (Figure 2.1.2). The bulk of selenium compounds in the crude will pass with the oil phase upon passing through the desalter. The crude contribution to the desalter Se effluent water will be negligible. It is the desalter wash water originating from the sour water stripper (SWS) that is responsible for the Se contribution to the desalter effluent. Most of the organoselenides convert to hydrogen selenide, particularly in the hydrotreaters. In the SWS, thiocyanates react with hydrogen selenide to create selenocyanates, which are the predominant species remaining in the water in the SWS bottoms. Most of any unreacted hydrogen selenide in the SWS goes overhead as a vapour to the sulfur recovery unit. Some elemental Se formed by high temperature breakdown of other species that stay with the heavier oil fractions upstream, particularly in the vacuum unit, is removed from the refinery via asphalt manufacturing.

It is a well-established process, but produces high amounts of sludge to be disposed and it is costly due to the use of hydrogen peroxide. The established biological process comprises of biological anoxic Se reduction-biomass adsorption (Figure 2.1.3) where a microbial community is applied to adsorb the selenite ions. However, complete biological reduction of selenium oxyanions, using the organic pollutants of the effluents is an alternative to solve this problem.

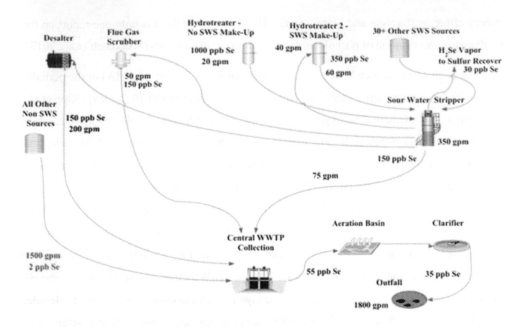

Figure 2.1.1 Fate of selenium in oil-refinery (Kujawski et al., 2014)

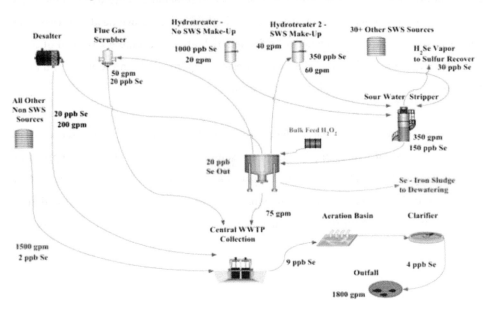

Figure 2.1.2 Iron Selenite Co-Precipitation - Se removal process. (Kujawski et al., 2014)

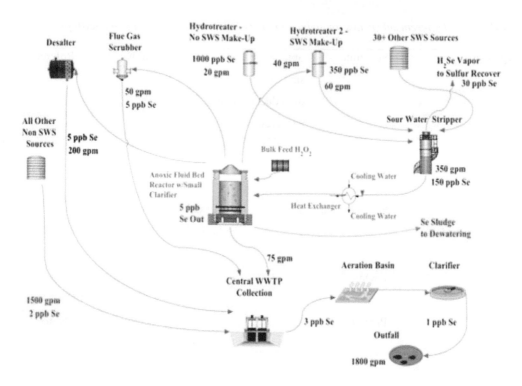

Figure 2.1.3 Biological anoxic Se Removal Process based on Se reduction and biomass adsorption (Kujawski et al., 2014)

In the field of biological remediation, the carbon substrates used for selenium removal to date are mainly organic carbon compounds like lactate (Tan et al., 2017), glucose (Espinosa-Ortiz et al., 2015) or Luria-Bertani broth (LB) (Zhang et al., 2019). Werkeneh et al. (2017) showed also phenol can be used as electron donor by *Phanerochaete chrysosporium*. In this reduction process, Se(0) nanoparticles are sometimes produced (Lee et al., 2007), both aerobically (Dhanjal et al., 2010) and anaerobically (Wadgaonkar et al., 2018). Nano Se(0) has immense applications in the nanomedicinal field including as antioxidant, chemopreventive agent, anti-fungal and anti-protozoan agent (Hosnedlova et al., 2018). Thus, biological production of nano Se(0) is very important, due to the variety of nano sized Se(0) produced without simultaneous production of hazardous substance given in Table 2.1. A brief overview of carbon compounds used for biological reduction of selenium oxyanions with production of nano Se(0) is also given in Table 2.1

Table 2.1 : Different substrates for reduction of selenium oxyanions and nano Se(0) produced

Microorganisms	Substrate	Smallest size (nm) of reduced particles	Shape of reduced particles	Location	References
E.Coli K12	LB broth	24	Spherical	Extracellular	Dobias et al., 2011
*Shewenella oneidensis*MR-1	Fumarate	100	Spherical	Intracellular	Li et al., 2014
Bacillus mycoides	Nutrient agar	50	Spherical	Extracellular	Lampis et al., 2014
Geobacter sulfurreducens	-	40	Spherical	-	Fellowes et al., 2011
Delftia lacustris	Lactate	-	-	-	Wadgaonkar et al., 2019
Phanerochaete chrysosporium	Phenol	-	-	Intracellular	Werkeneh et al., 2017

A variety of microbes can reduce selenium oxyanions (Nancharaiah and Lens, 2015). These selenium reducing microorganisms use various selenium conversions like reducing toxic selenium oxyanions into non-toxic and insoluble Se through hydrogen selenide (H_2Se) or dimethylselenide (Me_2Se) and methylselenocysteine (El-Ramady et al., 2014).

In most microorganisms (archaea and eubacteria), compounds containing selenium are metabolized along two pathways: (1) dissimilatory reduction of selenium oxyanions to Se^0 and possibly further to selenides or (2) their direct incorporation into amino acids (e.g., SeCys) and then to selenoproteins (Nancharaiah and Lens 2015) . An in-depth review of the ecological role, mechanism and phylogenetic characterization of various selenium-reducing microorganisms and their role in biotechnological applications.

2.2 Bioremediation of CO/syngas in the gaseous effluents of oil refinery and production of biofuels (alcohols) and value-added chemicals

Syngas is a mixture of carbon monoxide (CO), carbon dioxide (CO_2) and hydrogen (H_2) and is produced from various sources including petroleum refining, steel mill and various industries. Syngas fermentation is slowly emerging as a potential alternative to fuel production. The pathway followed by the CO fermenting organisms is the Wood-Ljungdahl pathway (Figure 2.1) to form an array of products including volatile fatty acids, alcohols, alkanes and even polyhydroxyalkanoates (Revelles et al., 2016). The equation with the formation of different acids and alcohols from CO or syngas are given in Table 2.2.

Table 2.2: Equations with synthesis of acids and alcohols from CO/syngas and Gibbs free energy (Fernández-Naveira et al., 2017b)

Production of acetic acid and ethanol from CO/syngas	
$6CO + 3H_2O \longrightarrow C_2H_5OH + 4CO_2$	$\Delta G° = -217.4$ kJ mol^{-1}
$6H_2 + 2CO_2 \longrightarrow C_2H_5OH + 3H_2O$	$\Delta G°= -97.0$ kJ mol^{-1}
$2CO + 4H_2 \longrightarrow C_2H_5OH + H_2$	$\Delta G°= -137.1$ kJ mol^{-1}
$3CO + 3H_2 \longrightarrow C_2H_5OH + CO_2$	$\Delta G° = -157.2$ kJ mol^{-1}
$4CO + 2H_2O \longrightarrow CH_3COOH + 2CO_2$	$\Delta G° = -154.6$ kJ mol$^-$
$4H_2 + 2CO_2 \longrightarrow CH_3COOH + 2H_2O$	$\Delta G° = -74.3$ kJ mol^{-1}
$2CO + 2H_2 \longrightarrow CH_3COOH$	$\Delta G° = -114.5$ kJ mol^{-1}
Production of butyric acid and butanol from CO/syngas	
$12CO+5H_2O \longrightarrow C_4H_9OH + 8CO_2$	$\Delta G° = -486.4$ kJ mol^{-1}
$12H_2 + 4CO_2 \longrightarrow C_4H_9OH + 7H_2O$	$\Delta G° = -245.6$ kJ mol^{-1}
$6CO + 6H_2 \longrightarrow C_4H_9OH + 2CO_2 + H_2O$	$\Delta G° = -373$ kJ mol^{-1}
$4CO + 8H_2 \longrightarrow C_4H_9OH + 3H_2O$	$\Delta G° = -334$ kJ mol^{-1}
$10CO + 4H_2O \longrightarrow CH_3(CH_2)_2COOH + 6CO_2$	$\Delta G° = -420.8$ kJ mol^{-1}
$10H_2 + 4CO_2 \longrightarrow CH_3(CH_2)_2COOH + 6H_2O$	$\Delta G° = -420.8$ kJ mol$^-$
$6CO + 4H_2 \longrightarrow CH_3(CH_2)_2COOH + 2CO_2$	$\Delta G° = -220.2$ kJ mol^{-1}
Production of hexanol and hexanoic acid from CO/syngas	
$18CO + 7H_2O \longrightarrow C_6H_{13}OH + 12CO_2$	$\Delta G° = -753$ kJ mol^{-1}
$18H_2 + 6CO_2 \longrightarrow C_6H_{13}OH + 11H_2O$	$\Delta G° = -395$ kJ mol^{-1}
$6CO + 12H_2 \longrightarrow C_6H_{13}OH + 5H_2O$	$\Delta G° = -514$ kJ mol^{-1}
$16CO + 6H_2O \longrightarrow CH_3(CH_2)_4COOH + 10CO_2$	$\Delta G° = -656$ kJ mol^{-1}
$16H_2 + 6CO_2 \longrightarrow CH_3(CH_2)_4COOH + 10H_2O$	$\Delta G° = -341$ kJ mol^1
$10CO + 10H_2 \longrightarrow CH_3(CH_2)_4COOH + 4CO2$	$\Delta G° = -540$ kJ mol ^{1}

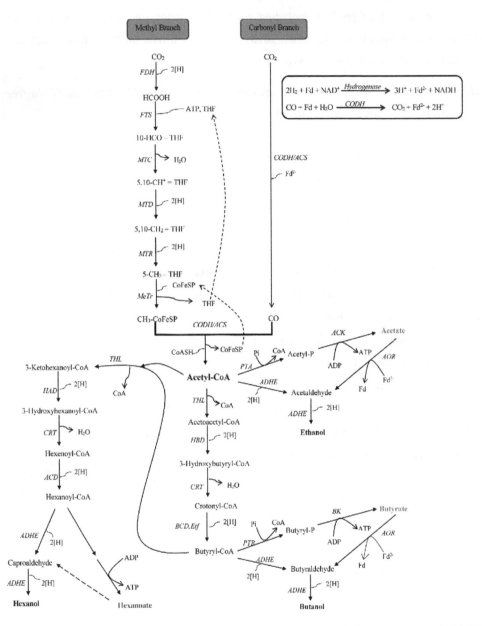

Figure 2.2: Wood-Ljungdahl pathway of CO fermentation (Fernández-Naveira et al., 2017b)

The organisms performing CO/syngas fermentaton mentioned in Table 2.3 are mostly belonging to *Clostridium* sp. with the exception of *Alkalibaculum bacchi* and *Moorella* sp. The hexanol producing strain from CO is *Clostridium carboxidivorans*.

Table 2.3: Microorganisms performing CO/syngas fermentation (Sun et al., 2019)

Microorganism	Growth pH	Growth temperature (°C)	Products
Alkalibaculum bacchi	6.5–10.5	15–40	Acetate, ethanol
Butyribacterium methylotrophicum	5.5–6.0	37	Acetate, ethanol, butyrate, butanol
Clostridium carboxidivorans	4.4–7.6	24–42	Acetate, ethanol, butyrate, butanol, caproate, hexanol
Clostridium ragsdalei	5.0–7.5	25–40	Acetate, ethanol, 2,3-butanediol
Clostridium autoethanogenum	4.5–6.5	20–44	Acetate, ethanol, 2,3-butanediol
Clostridium ljungdahlii	4.0–6.0	30–40	Acetate, ethanol, 2,3-butanediol, formic acid
Clostridium kluyveri	6.0–7.5	30	Butyrate, caproate, H_2
Clostridium drakei	4.6–7.8	18–42	Acetate, ethanol, butyrate, butanol
Clostridium scatologenes	4.6–8.0	18–42	Acetate, butyrate
Eubacterium limosum	7.0–7.2	38–39	Acetate, butyrate
Sporomusa ovata	5.0–8.1	15–45	Acetate, ethanol
Moorella thermoacetica	6.5	55	Acetate
Moorella thermoautotrophica	5.7	36–70	Acetate
Moorella stamsii	5.7–8.0	50–70	Acetate, H_2
Moorella glycerini	6.3–6.5	58	Not reported
Moorella perchloratireducens	5.5–8.0	37	Acetate

Peptostreptococcus productus	7.0	37	Acetate
Alkalibaculum bacchi and *Clostridium propionicum*	6.0–8.0	37	Acetate, ethanol, propionate, propanol, butyrate, butanol, hexanol
Clostridium autoethanogenum and Clostridium kluyveri	5.5–6.5	37	Acetate, ethanol, butyrate, butanol, caproate, hexanol
Clostridium ljungdahlii and Clostridium kluyveri	5.7–6.4	35	Acetate, ethanol, butyrate, butanol, caproate, hexanol, 2,3-butanediol, octanol
Genetically modified			
Clostridium ljungdahlii B6	Not reported	Not reported	Butyrate, ethanol, acetate
Acetobacterium woodii [pMTL84151_act$_{thlA}$]	7.0	30	Acetone, acetate
Clostridium ljungdahlii(pSOBP$_{ptb}$)	Not reported	37	Butanol, butyrate, ethanol, acetate
Clostridium autoethanogenum	Not reported	37	Fatty acid ethyl ester, fatty acid butyl ester
Clostridium ljungdahlii	Not reported	37	Isoprene
Clostridium autoethanogenum	Not reported	37	3-Hydroxypropionate (3-HP)
Clostridium autoethanogenum PHB	5.0	37	Poly-3-hydroxybutyrate (PHB)
Clostridium coskatii[p83_tcb]	Not reported	37	3-Hydroxybutyrate (3-HB)

22

Clostridium carboxidivorans is perhaps the most researched organism producing caproate and hexanol (Fernández-Naveira et al., 2019), followed by *Clostridium autoethanogenum* for 2, 3 butanediol production (Abubacker et al., 2015) and *Clostridium ragsdalei* (Saxena and tanner 2011). A co-culture of *Clostridium kluyveri* and *Clostridium ljungdahlii* can produce a range of butanol, ethanol, hexanol and octanol co-metabolism (Richter et al., 2016). A co-culture of *Clostridium kluyveri* and *Clostridium autoethanogenum* can produce butanol, ethanol and hexanol (Diender et al., 2016).

2.3 Key parameters for alcohol production by syngas fermentation

In the era of 3rd generation of biofuels, syngas/CO fermentation is of utmost importance as it produces desired biofuels, among others alcohols, from CO/syngas. Research is in full swing to find the commercial production of these compounds to have a better understanding of the regulatory factors of CO/syngas fermentation along with the external parameters like pH, temperature, partial pressure of the gases and impurities present in industrial syngas. Various trace metals propel the cellular machinery to produce the desired fermentation product (Saxena and Tanner 2011). Particularly in case of alcohol production, certain factors like pH and important metal co-factors (like tungsten and selenium) play a crucial role in the process of solventogenesis, i.e., production of alcohols (Saxena and Tanner 2011; Abubacker et al., 2015, Fernández-Naveira et al., 2019).

2.3.1 Role of pH

The pH plays an important role in alcohol production. When acetic acid is released outside the cell, being a weak lipophilic acid it permeates inside the cell creating a low pH condition (Abubackar et al., 2012). Permeation occurs due to low molecular weight of the acids. A low extracellular pH would further impose a stress on the cell leading to the switching from acetogenesis to solventogenesis. Hence, the intracellular pH is increased for resuming the regular cell metabolism (Abubackar et al., 2012). However, a very low pH (below 4) would completely inhibit the cell activity. Recently, production of ethanol has been found to be favored at low fermentation pH in mixed cultures, but it stops production of longer chain alcohols like hexanol (Ganigué et al., 2016). Altering the pH throughout the fermentation of

Clostridium autoethanogenum (Abubackar et al., 2016) has been found to improve the alcohol productivity. Production of acids at higher pH of 6.2 and conversion of these acids to alcohols have also been observed after changing the pH to 4.9 consecutively by pure cultures like *Clostridium carboxidivorans* (Fernández-Naveira et al., 2017a).

2.3.2. Tungsten (W): heaviest essential element in anaerobic microbiology

Tungsten is the heaviest element involved in the metabolism of biological systems and it is an essential constituent of almost all the enzymes belonging to the aldehyde reductase family (AOR) (Andreesen et al., 2008). The first evidence of tungsten as an important cofactor was shown by formate dehydrogenase (FDH) activity of the Wood-Ljungdahl pathway of *Moorella thermoaceticum* in 1979 (Andreesen et al., 2008). In *Clostridium ragsdalei*, growing on syngas as the substrate, indicated that the presence of tungsten, at a concentration of 0.681 μM (in the form of tungstate), yielded an ethanol production of 35.73 mM, which was enhanced to 72.3 mM upon increasing the tungsten concentration to 6.81 μM (Saxena and Tanner, 2011). In that study, it was suggested that the presence of both selenium and tungsten in the medium decreases the activity of FDH in *C. ragsdalei* compared to media containing either tungsten or selenium, but increases alcohol formation, i.e. solventogenesis.

Aldehyde Ferredoxin Oxidoreductase (AFOR) catalyses acetic acid reduction to acetaldehyde (Abubacker et al., 2015). The only known AFOR with structural elucidation is extracted from hyper thermophilic archaeon *Pyrococcus furiosus*. The enzyme is a homodimer with 4–5 Fe atoms and 1 W atom per molecule (Kletzin and Adams, 1996). Tungsten increases the butanediol/acetic acid ratio similarly as the ethanol/acetic acid ratio. The ethanol/acetic acid ratio increased by 5 to 7 times after addition of either low or high concentration of tungsten (, while there is 20% increase in butanediol/acetic acid ratio increased after addition of tungsten. For both ethanol and 2, 3-butanediol, their relative concentration is the highest (compared to acetic acid) in medium supplemented only with tungsten (no vitamin-solution and no selenium) followed by medium containing no tungsten nor selenium and then medium supplemented only selenium (no tungsten) (Abubacker et al., 2015). In the main branch of Wood-Ljungdahl pathway and later in the branch of 2, 3-butanediol production, using pyruvate: ferredoxin oxidoreductase (PFOR), acetyl-CoA with CO_2 is converted to pyruvate. Pyruvate gets reduced by acetolactate synthase and acetolactate decarboxylase to acetoin and then later to 2, 3-butanediol using 2, 3-butanediol dehydrogenase (2,3BDH) (Köpke et al., 2011).

Increasing the tungstate concentration by 10 fold from 0.1 μm, the bacterium *Spormusa ovuta,* produces 4.4 fold more acetate from 32.0 (±1.7) to 141.2 (±56.6) mmol m²/day and 1.5 ± (0.5) mmol ethanol with a production rate of 4.0 (± 1.2) mmol m^{-2} day^{-1} (48.0 ± 14.6 μM day^{-1}) from day 1 to 4 in a microbial electrosynthetic system driving current from a cathode source with CO_2 as the sole carbon substrate (Ammam et al., 2016). Electron recovery in both acetate and ethanol with tenfold concentration of tungstate was 87.6 (±6.5%). The enhanced production of acetate and ethanol was attributed to the upregulation of the tungstate containing FDH enzyme of the Wood-Ljungdahl pathway substantiated by the abundant transcripts of six AOR genes. External addition of fatty acids like 1-propanoic acid and 1-butyric acid yielded 1-propanol (1.3 fold higher in tenfold tungstate concentration) and 1-butanol (1.6 fold higher in tenfold tungstate concentration) validated by the higher alcohol dehydrogenase (ADH) activity. The enhanced capacity of metabolic reduction by *S. ovuta* was found to be induced by tungsten and it unfolds novel bio-production of higher alcohols in an electricity driven system.

2.3.3 Selenium: a cofactor of solventogenic enzymes

Selenium has always been mentioned as a component of the fermentation medium for syngas fermentation (Saxena and Tanner 2011, Liu et al., 2012, Fernández-Naveira et al., 2017a, b). Only in case of *Clostridium ragsdaleii* (Saxena and Tanner, 2011), selenium (in the form of sodium selenate) was found to have a positive effect on ethanol production. 38.32 mM ethanol (1.76 g/L) increased to 54.35 mM ethanol (2.5g/L) when the selenium (in the form of selenate) concentration increased from 1.06 μM to 5.03 μM. But in case of *Clostridium autoethanogenum*, elimination of selenium improved production of alcohol by 25%. (Abubacker et al., 2015). This shows the selectivity of the effect of the same trace element among two similar species of CO fermenting bacteria. To the best of our knowledge, these are the only studies available on the effect of selenium. The known enzymatic studies have revealed that the purified tetrameric enzyme of NADP dependent formate dehydrogenase of *Clostridium thermoaceticum* (Gollin et al., 1998) contains per molecule two tungsten, two selenium, 36 iron, and about 50 sulfur atoms. The Se is in the form of seleno-cysteine situated in the larger Gt subunit of the enzyme. Further research is necessary for finding the actual role of selenium in the CO/syngas fermentation pathway.

2.4 Alcohol recovery from fermentation broth

2.4.1 Definition of alcohol and importance of butanol as fuel

Alcohols are compounds containing an OH (hydroxyl) group and can also be produced by anaerobic fermentation. There are several processes to recover alcohols, the product of syngas fermentation, from the fermentation broth (Figure 2.3), including fractional distillation for ethanol (Blum et al., 2010; Tgarguifa et al., 2017), extractive distillation, (Errico et al., 2013) and pressure swing adsorption (Simo et al., 2008). Sometimes extractive distillation is also applied to recover ethanol (Errico et al., 2013). For butanol recovery from typical acetone-butanol-ethanol fermentation broth (Figure 2.3), solvent extraction, perstraction, vacuum evaporation, flash fermentation and gas-stripping methods can be used (Jiménez-Bonilla et al., 2018).

Butanol as a fuel or fuel additive is quite advantageous (Liu et al., 2014), but fermentative butanol production, due to low concentration is not economically viable, due high requirement of energy for the process. In view of industrial production, regular distillation, vacuum distillation, flash fermentation and gas stripping do not show significant progress and current research. Synthesis of new materials is continuously inducing the growth of membrane-associated technologies (pervaporation and perstraction). But the most energy efficient way is adsorption (sometimes it is followed by absorption) of selected compound (Qureshi et al., 2005), also due to the bacterial growth in the adsorption materials.

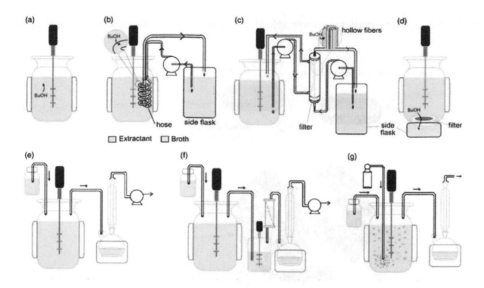

Figure 2.3: Different processes of butanol recovery systems: (a) Direct extract-broth contact process for regular solvent extraction; (b) "Tube"-type perstraction: ; butanol migrates from the broth (in contact with the external of the hose) to the extractant (inside of the hose)extractant is pumped through the fermentation flask with no direct contact with the broth, (c) Fiber filter perstraction: the extractant is driven inside of the hollow membrane, broth is pumped through the shell and returned back to the bioreactor; (d) Perstraction by membrane filter: a circular filter separates the broth components and the extractant; butanol exchange takes place through the filter; (e) vacuum evaporation; (f) flash fermentation; and (g) gas stripping. (Jiménez-Bonilla et al., 2018).

2.4.2 Adsorption

The phenomenon utilized for separation of compounds from solutions is called adsorption. It occurs when a solid phase is introduced to a system of either liquid or gas components and this solid phase then adsorbs one of the components to its matrix. Mass transfer occurs in adsorption in which the adsorbate particles are concentrated on the adsorbent (solid phase) surface and thus it is a surface phenomenon. At low temperature, adsorption is spontaneous with negative Gibbs free energy ($\Delta G^{\circ} < 0$). Physisorption and chemisorption are two kinds of adsorption depending on the mechanism. Physical forces like weak Van der Waals forces bind the particles of adsorbate and adsorbent in physisorption. In chemisorption the interaction between the adsorbate and adsorbent particles is more selective and stronger, like the valence forces (Figure 2.4).

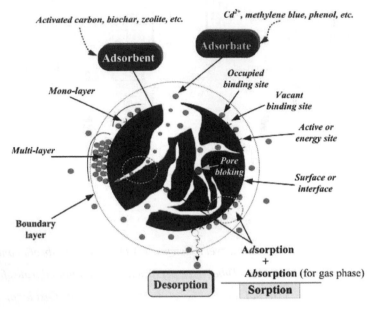

Figure 2.4 Different terms used to explain adsorption phenomena (Tran et al., 2017)

In case of alcohols, butanol being an ideal biofuel, commercial products like high silica zeolites CBV28014 have been found to adsorb butanol selective over water (Oudshoorn et al., 2009). From a model syngas fermentation broth, two commercial resins, Dowex Optipore™ L-493 and Diaion®HP-20 along with granular activated carbon Norit®1240w GAC were used to find the capability of adsorptive recovery of acetic acid, ethanol and butanol (Sadrimajd et al., 2019). This study showed that Dowex Optipore™ L-493 was more selective towards butanol and a lower adsorbent dose showed higher adsorption capacity (0.3g/20 mL).

2.4.3 Absorption

Absorption is a phenomenon where a liquid or gas is taken up by a solid and retains it within its core. The difference between adsorption and absorption is that adsorption is a surface phenomenon and no morphological changes take place in the inside of the absorbent and, hence can retain it for a long period of time. In case of alcohols, certain organic compounds called organic gels are able to absorb alcohols (Kabiri et al., 2011a, b), at a very high concentration of alcohols in the absence of water. It is also necessary to synthesize and find the applicability of organic alcogels in retaining the alcohols produced at low concentrations (0-5 g/L) in fermentation broths.

2.5 Conclusion

The current scenario of biovalorisation of the liquid and gaseous effluents of oil refinery and petrochemical industry is yet underdeveloped. The liquid effluents containing the phenolic effluents and toxic metalloids offer paramount scope of bio detoxification and simultaneous recovery of nanoparticles. On the other hand, the energy quotient and economical aspects of bioconversion of CO/syngas, the gaseous effluent of oil refinery, can be a potential tool of producing biofuels. Further investigations of the microbiological process and endeveours for upscaling are required in this field of study.

References

Abdelwahab, O., Amin, N.K., El-Ashtoukhy, E.Z. Electrochemical removal of phenol from oil refinery wastewater. J. Hazard. Mater. 163 (2009): 711-716.

Abubackar, H.N., Veiga, M.C., Kennes, C. Biological conversion of carbon monoxide to ethanol: effect of pH, gas pressure, reducing agent and yeast extract. Bioresour. Technol. 114 (2012): 518-522.

Ammam, F., Tremblay, P.L., Lizak, D.M., Zhang, T. Effect of tungstate on acetate and ethanol production by the electrosynthetic bacterium Spormusa ovuta. Biotechnol. Biofuels. 9 (2016): 163-172.

Andreesen, J.R., Makdessi, K. Tungsten, the surprisingly positively acting heavy metal element for prokaryotes. Ann. N. Y. Acad. Sci. 1125 (2008): 215-229.

Al Hashemi, W., Maraqa, M.A., Rao, M.V. and Hossain, M.M., 2015. Characterization and removal of phenolic compounds from condensate-oil refinery wastewater. Desalin. Water Treat. 54 (2015): 660-671.

Blum, S.R., Kaiser, B. Buss-SMS-Canzler GmbH and 2S-sophisticated systems Ltd, 2010. Distill. Method. U.S. Patent 7,744,727.

Chasteen TG, Bentley R. Biomethylation of selenium and tellurium. Chem. Rev. 103 (2003): 1-25.

Diender, M., Stams, A.J., Sousa, D.Z. Production of medium-chain fatty acids and higher alcohols by a synthetic co-culture grown on carbon monoxide or syngas. Biotechnol. biofuels. 9(2016): p.82.

Dhanjal, S., Cameotra, S.S. Aerobic biogenesis of selenium nanospheres by *Bacillus cereus* isolated from coalmine soil. Microb. Cell Fact. 9 (2010): 52-62.

Dhillon K.S., Dhillon, S.K. Distribution and management of seleniferous soils.In: Sparks D. (ed) Adv. in Agro. Academic Press (2003): 119–184.

Dobias, J., Suvorova, E.I., Bernier-Latmani, R. Role of proteins in controlling selenium nanoparticle size. Nanotechnol. 22(2011): 195605-195614

El-Naas, M.H., Al-Zuhair, S., Alhaija, M.A. Removal of phenol from petroleum refinery wastewater through adsorption on date-pit activated carbon. Chem. Eng. J. 162 (2010): 997-1005.

El-Ramady, H., Abdalla, N., Alshaal, T., Domokos-Szabolcsy, E., Elhawat, N., Prokisch, J., Sztrik, A., Fari, M., El-Marsafawy, S. and Shams, M.S. Selenium in soils under climate change, implication for human health. Environ. Chem. Lett. 13 (2014): 1-19.

Errico, M., Rong, B.G., Tola, G., Spano, M. Optimal synthesis of distillation systems for bioethanol separation. Part 1: Extractive distillation with simple columns. Ind. Eng. Chem. Res. 52 (2013): 1612-1619.

Espinosa-Ortiz, E.J., Gonzalez-Gil, G., Saikaly, P.E., Hullebusch, E.Dv. , Lens, P.N.L. Effects of selenium oxyanions on the white-rot fungus Phanerochaete chrysosporium. Appl. Microbiol. Biotechnol. 99(2015): 2405-2418.

Fellowes, J.W., Pattrick, R.A.D., Green, D.I., Dent, A., Lloyd, J.R., Pearce, C.I. Use of biogenic and abiotic elemental selenium nanospheres to sequester elemental mercury released from mercury contaminated museum specimens. J. Hazard. Mater. 189 (2011): 660-669.

Fernández-Naveira, Á., Veiga, M.C., Kennes, C. Effect of pH control on the anaerobic H-B-E fermentation of syngas in bioreactors. J. Chem. Technol. Biotechnol. 92 (2017a): 1178-1185.

Fernández-Naveira, Á., Veiga, M.C., Kennes, C. H-B-E (hexanol-butanol-ethanol) fermentation for the production of higher alcohols from syngas/waste gas. J. Chem. Technol. Biotechnol. 92 (2017b): 712-731.

Ganigué, R., Sanchez-Paredes, P., Baneras, L., Colprim, J. Low fermentation pH is a trigger to alcohol production, but a killer to chain elongation. Frontiers in microbiology, 7 (2016): 702.

Gollin, D., Li, X.L., Liu, S.M., Davies, E.T., Ljungdahl, L.G. Acetogenesis and the primary structure of the NADP-dependent formate dehydrogenase of Clostridium thermoaceticum, a tungsten-selenium-iron protein. Stud. Surf. Sci. Catal. 114(1998): 303-308.

Gupta, A., Balomajumder, C. Simultaneous removal of Cr (VI) and phenol from binary solution using Bacillus sp. immobilized onto tea waste biomass. J. Water Process Eng. 6 (2015): 1-10.

Hernández-Francisco, E., Peral, J., Blanco-Jerez, L.M. Removal of phenolic compounds from oil refinery wastewater by electrocoagulation and Fenton/photo-Fenton processes. J. Water Proc. Eng. 19 (2017): 96-100.

Hosnedlova, B., Kepinska, M., Skalickova, S., Fernandez, C., Ruttkay-Nedecky, B., Peng, Q., Baron, M., Melcova, M., Opatrilova, R., Zidkova, J., Bjørklund, G. Nano-selenium and its nanomedicine applications: a critical review. Int. J. Nanomedicine. 13 (2018). 2107.

Hurst, K.M., Lewis, R.S. Carbon monoxide partial pressure effects on the metabolic process of syngas fermentation. Biochem. Eng. J. 48 (2010): 159-165.

Jiménez-Bonilla, P., Wang, Y. In situ biobutanol recovery from clostridial fermentations: a critical review. Crit. Rev. Biotechnol. 38 (2018): 469-482.

Kabiri, K., Lashani, S., Zohuriaan-Mehr, M.J., Kheirabadi, M. Super alcohol-absorbent gels of sulfonic acid-contained poly (acrylic acid). J. Polym. Res. 18(2011a): 449-458.

Kabiri, K., Azizi, A., Zohuriaan-Mehr, M.J., Marandi, G.B. and Bouhendi, H. Poly (acrylic acid–sodium styrene sulfonate) organogels: preparation, characterization, and alcohol superabsorbency. J. Appl. Polym. Sci. 119(2011b): 2759-2769.

Kavitha, V., Palanivelu, K. The role of ferrous ion in Fenton and photo-Fenton processes for the degradation of phenol. Chemosphere, 55 (2004): 1235-1243.

Kipp, A.P., Strohm, D., Brigelius-Flohé, R., Schomburg, L., Bechthold, A., Leschik-Bonnet, E., Heseker, H., DGE, G.N.S. Revised reference values for selenium intake. J Trace. Elem. Med. Biol. 32 (2015): 195-199.

Kletzin, A., Adams, M.W. Tungsten in biological systems. FEMS Microbiol. Rev. 18 (1996): 5-63.

Köpke, M., Mihalcea, C., Liew, F., Tizard, J.H., Ali, M.S., Conolly, J.J., Al-Sinawi, B., Simpson, S.D. 2, 3-Butanediol production by acetogenic bacteria, an alternative route to chemical synthesis, using industrial waste gas. Appl. Environ. Microbiol. 77 (2011), 5467-5475.

Kujawski D. (2014, April 24). Retrieved from URL https://www.waterworld.com/industrial/article/16211131/the-scoop-on-selenium-exploring-sources-fate-and-transport-of-se-in-oil-refining

Lawson S., Macy J.M. Bioremediation of selenite in oil refinery wastewater. Appl. Microbiol. Biotechnol. 43(1995): 762-765.

Lampis, S., Zonaro, E., Bertolini, C., Bernardi, P., Butler, C.S., Vallini, G. Delayed formation of zero-valent selenium nanoparticles by Bacillus mycoides SeITE01 as a consequence of selenite reduction under aerobic conditions. Microbiol. Cell. Fact. 13(2014): 1-14.

Lee, J.H., Han, J., Choi, H., Hur, H.G. Effects of temperature and dissolved oxygen on Se (IV) removal and Se (0) precipitation by Shewanella sp. HN-41. Chemosphere, 68(2007): 1898-1905.

Li, D.B., Cheng, Y.Y., Wu, C., Li, W.W., Li, N., Yang, Z.C., Tong, Z.H., Yu, H.Q. Selenite reduction by Shewanella oneidensis MR-1 is mediated by fumarate reductase in periplasm. Sci. Rep. 4(2014): 3735-3741.

Liu, K., Atiyeh, H.K., Tanner, R.S., Wilkins, M.R., Huhnke, R.L. Fermentative production of ethanol from syngas using novel moderately alkaliphilic strains of *Alkalibaculum bacchi*. Bioresour. Technol. 104(2012): 336-341.

Liu, H., Wang, X., Zheng, Z., Gu, J., Wang, H., Yao, M. Experimental and simulation investigation of the combustion characteristics and emissions using n-butanol/biodiesel dual-fuel injection on a diesel engine. Energy, 74 (2014): 741-752.

Nacharaiah, Y.V., Lens, P.N.L. The ecology and biotechnology of selenium respiring bacteria. Microbiol. Mol. Bio. Rev 79 (2015): 61-80

Ojumu, T.V., Bello, O.O., Sonibare, J.A., Solomon, B.O. Evaluation of microbial systems for bioremediation of petroleum refinery effluents in Nigeria. Afr. J. Biotechnol. 4 (2005): 31-35.

Ontañon, O.M., González, P.S., Barros, G.G., Agostini, E. Improvement of simultaneous Cr (VI) and phenol removal by an immobilized bacterial consortium and characterization of biodegradation products. New Biotechnol. 37 (2017): 172-179.

Oudshoorn, A., van der Wielen, L.A., Straathof, A.J. Adsorption equilibria of bio-based butanol solutions using zeolite. Biochem. Eng. J. 48 (2009): 99-103.

Sadrimajd, P., Rene, E.R., Lens, P.N.L. Adsorptive recovery of alcohols from a model syngas fermentation broth. Fuel, 254 (2019), 115590-115597.

Pradeep, N.V., Anupama, S., Navya, K., Shalini, H.N., Idris, M., Hampannavar, U.S. Biological removal of phenol from wastewaters: a mini review. Appl. Water Sci. 5 (2015): 105-112.

Qureshi, N., Hughes, S., Maddox, I.S, Cotta, M.A. Energy efficient recovery of butanol from model solutions and fermentation broth by adsorption. Bioproc. Biosyst. Eng. 27 (2005): 215-222.

Rehman, K., Imran, A., Amin, I., Afzal, M. Enhancement of oil field-produced wastewater remediation by bacterially-augmented floating treatment wetlands. Chemosphere, 217 (2019): 576-583.

Revelles, O., Tarazona, N., García, J.L., Prieto, M.A. Carbon roadmap from syngas to polyhydroxyalkanoates in Rhodospirillum rubrum. Environ. Microbiol. 18 (2016): 708-720.

Richter, H., Molitor, B., Diender, M., Sousa, D.Z., Angenent L.T., 2016. A narrow pH range supports butanol, hexanol, and octanol production from syngas in a continuous co-culture of Clostridium ljungdahlii and Clostridium kluyveri with in-line product extraction. Frontiers Microbiol. *7* (2016): 1773.

Saxena, J., Tanner, R.S. Effect of trace metals on ethanol production from synthesis gas by the ethanologenic acetogen, *Clostridium ragsdalei*. J. Ind. Microbiol. Biotechnol. 38 (2011): 513-521.

Simo, M., Brown, C.J., Hlavacek, V. Simulation of pressure swing adsorption in fuel ethanol production process. , Comput. Chem. Eng. 32 (2008): 1635-1649

Sun, X., Atiyeh, H.K., Huhnke, R.L., Tanner, R.S. Syngas fermentation process development for production of biofuels and chemicals: A review. Bioresour. Technol. Rep (2019): 100279-100328.

Tan, L.C., Nancharaiah, Y.V., van Hullebusch, E.D., Lens, P.N.L. Effect of elevated nitrate and sulphate concentrations on selenate removal by mesophilic anaerobic granular sludge bed reactors. Env. Sci.: Wat-Res. Technol. 4 (2018): 303-314.

Tgarguifa, A., Abderafi, S., Bounahmidi, T. Modeling and optimization of distillation to produce bioethanol. Energy Procedia, 139 (2017): 43-48.

Tran, H.N., You, S.J., Hosseini-Bandegharaei, A., Chao, H.P. Mistakes and inconsistencies regarding adsorption of contaminants from aqueous solutions: a critical review. Water Res. 120 (2017): 88-116.

Wadgaonkar, S.L., Mal, J., Nancharaiah, Y.V., Maheshwari, N.O., Esposito, G. Lens, P.N.L. Formation of Se(0), Te(0), and Se(0)–Te (0) nanostructures during simultaneous bioreduction of selenite and tellurite in a UASB reactor. Appl. Microbiol. Biotechnol. 102 (2018): 2899-2911.

Wadgaonkar, S.L., Nancharaiah, Y.V., Jacob, C., Esposito, G., Lens, P.N.L. Microbial transformation of Se oxyanions in cultures of *Delftia lacustris* grown under aerobic conditions. J. Microbiol. 57 (2019): 362-371.

Werkeneh, A.A., Rene, E.R. Lens, P.N.L. Simultaneous removal of selenite and phenol from wastewater in an upflow fungal pellet bioreactor. J. Chem. Technol. Biotechnol. 93 (2018): 1003-1011.

Zhang, J., Wang, Y., Shao, Z., Li, J., Zan, S., Zhou, S. and Yang, R. Two selenium tolerant *Lysinibacillus* sp. strains are capable of reducing selenite to elemental Se efficiently under aerobic conditions. J. Environ. Sci. 77 (2019): 238-249.

Chapter 3

Aerobic fungal-bacterial co-culture to detoxify phenolic effluents and concomitant reduction of selenite ions of oil-refinery containing selenite ions

This chapter has been published as:

Chakraborty S, Rene, E.R., Lens, P.N.L. Reduction of selenite to elemental Se(0) with simultaneous degradation of phenol by co-cultures of *Phanerochaete chrysosporium* and *Delftia lacustris*. *Journal of Microbiology*, 57(2019): 738-747.

Abstract

The simultaneous removal of phenol and selenite from synthetic wastewater was investigated by adopting two different co-culturing techniques using the fungus *Phanerochaete chrysosporium* and the bacterium *Delftia lacustris*. Separately grown biomass of the fungus and the bacterium (suspended co-culture) was incubated with different concentrations of phenol (0-1200 mg/L) and selenite (10 mg/L). The selenite ions were biologically reduced to extracellular Se(0) nanoparticles (3.58 nm diameter) with the simultaneous degradation of up to 800 mg/L of phenol. Upon growing the fungus and the bacterium together using an attached growth co-culture, the bacterium grew as a biofilm onto the fungus. The extracellularly produced Se(0) in the attached growth co-culture had a minimum diameter of 58.5 nm. This co-culture was able to degrade completely 50 mg/L phenol, but was completely inhibited at a phenol concentration of 200 mg/L.

Keywords: *Phanerochaete chrysosporium*, *Delftia lacustris*, phenol, selenite, Se(0) nanoparticles

3.1 Introduction

Petroleum and oil refineries discharge large quantities of phenol and polyaromatic hydrocarbons which have toxic, mutagenic and carcinogenic effects on flora and fauna, even at very low concentrations (Abdelwahab et al., 2009). Rapid industrialization has led to the gradual accumulation of these recalcitrant phenolic compounds in water bodies (Pradeep et al., 2015), which are often found together with certain metalloids. For instance, selenium is a metalloid frequently found in the form of toxic selenite in oil refinery wastewaters (Lawson and Marcy 1995). There is a marginal difference in the selenium concentrations for being an essential nutrient or a toxic element (Tan et al., 2016). In some refinery wastewater, selenium concentrations ranging from 0.1 to 3.7 mg/L can be found (Nurdogan et al., 2012), while phenol concentrations can be as high as 230 mg/L (Almendariz et al., 2005; Hashemi et al., 2015). The toxic selenites in the presence of phenolic pollutants pose a great challenge, due to the high cumulative toxicity it exerts on all life-forms. Thus, the simultaneous biological removal of selenite and phenolic pollutants from petrochemical wastewater is of utmost importance.

Most of the previous studies have mainly focussed on bacterial anaerobic reduction, except few studies which focus on aerobic bacterial reduction (Presentato et al., 2017). The detoxification mechanism of selenite ions and the selenite tolerance of fungi has also been reported in the literature (Gadd 2007). The simultaneous reduction of selenite and degradation of phenol by fungal and bacterial cells has thus far, not been investigated. Fungal-bacterial co-cultures can mineralize the pollutants more efficiently because the substrates toxic to bacteria can be degraded by the fungus and vice versa (Cheng et al., 2017). Biological aerobic reduction of the selenium oxyanions coupled with the detoxification of these recalcitrant organic pollutants can lay the foundation of an inexpensive, energy-efficient, dual remediation process of hydrocarbons and selenium oxyanions present in the effluents of the petrochemical industry.

P. chrysosporium is a well-studied white-rot fungus capable of detoxifying many polyaromatic hydrocarbons and phenolic compounds, mainly due to its ability to produce lignolytic enzymes during the secondary metabolic growth phase of the fungus. This white-rot fungus can reduce up to 10 mg/L selenite using glucose as the electron donor (Espinoza-Ortiz et al., 2015). The phenol degradation capacity in the presence of selenite by *P. chrysosporium* has been reported (Werkeneh et al., 2017). *Delftia lacustris* is a recently isolated bacterium which can reduce selenium oxyanions using lactate as the carbon source (Wadgaonkar et al., 2019).

A fungal-bacterial co-culture may thus stimulate the growth of the bacterium in the presence of phenol and enhance the reduction efficiency of selenite ions. The main objectives of this work were: (i) to investigate and compare the efficiencies of the simultaneous degradation of phenol and reduction of selenite ions by employing two different modes of growth of the fungal-bacterial co-culture, (ii) to identify the mechanisms involved in this type of reduction of selenium oxyanions coupled to the simultaneous degradation of phenol mediated by a fungal microbial consortium of *P. chrysosporium* and *D. lacustris*, and (iii) to characterize the Se(0) produced by both co-cultures.

3.2 Materials and methods

3.2.1 Microorganisms and growth of fungal-bacterial co-cultures

The white-rot fungus *P. chrysosporium* (MTCC187) was obtained from the Institute of Microbial Technology (Chandigarh, India) and was maintained on malt extract agar plates, at 37°C for 3 days and the spores were transferred to a glucose containing (10 g/L) medium (Espinosa-Ortiz et al. 2015) for the pelletization of the fungus. The fungal biomass was incubated at 30°C, for 2 days, at a pH of 4.5 under continous stirring at 150 rpm. The composition of the medium, except the carbon sources, was similar to Espinoza-Ortiz et al. (2015). Simultaneously, *D. lacustris* (NCBI MH158542) was separately grown in a lactate rich medium to obtain an actively growing bacterial biomass (Jørgensen et al., 2009). The fungal and bacterial biomass were thus grown separately. This type of co-culture is hereafter referred as "suspended co-culture". In order to determine the biodegradation capability of the fungus and bacterium separately, *D. lacustris* was incubated with phenol at concentrations ranging from 0 to 100 mg/L, in the presence of 0 and 10 mg/L of selenite.

When *D. lacustris and P. chrysosporium* were incubated together, the bacterium grew as a biofilm over the fungal cells and this type of co-culture is hereafter referred as "attached co-culture". Table 1 overviews the different carbon substrates used for establishing the co-culture. The 3-day old spore suspension of *P. chrysosporium* and the 2-day old active *D. lacustris* were

40

mixed and allowed to grow together in glucose (5 g/L) and lactate (5 g/L) containing medium, for 2 days, at a pH of 6.5. The composition of the medium used for the growth of the attached co-culture, except the carbon sources, was similar to those reported by Espinoza-Ortiz et al. containing 2 g/L KH_2PO_4, 0.5 g/L $MgSO_4.7H_2O$, 0.1 g/L NH_4Cl, 0.1 g/L $CaCl_2$, 0.001 g/L thiamine and 5ml/L trace element solution (2015).

3.2.2 Batch experiments

All experiments were carried out in airtight Erlenmeyer flasks (250 mL) with a working volume of 100 mL. In the suspended co-culture system, the fungal pellets were transferred to the phenol-rich medium after 2 days of incubation, wherein the phenol concentration was varied from 0 to 1200 mg/L and 100 µL of the active *D. lacustris* culture was added to the medium and maintained at a pH of 6.5, temperature of 30°C and under agitation at 180 rpm. The selenite concentration was maintained constant at 10 mg/L. The control results for phenol degradation by *P. chrysosporium* solely was compared and standardised with respect to a previous work by Werkeneh et al. (2017).

In the attached co-culture system, the fungal-bacterial biomass was transferred aseptically to the medium wherein the phenol concentrations were varied from 0 to 200 mg/L, while the selenite concentration was maintained constant at 10 mg/L, at pH 6.5, 30°C and agitation at 180 rpm agitation. About 1 mL of the sample from both co-culture systems were collected at an interval of 12 h to study the removal of phenol, growth rate of the microorganisms, the selenite reduction efficiency and the production of any detectable carbon compounds.

3.2.3 Analytical methods

3.2.3.1 Biomass analysis

In the suspended co-culture system, the biomass concentration in the liquid phase was determined gravimetrically by centrifuging 1 mL of the liquid broth and determining the dry weight of the pellet. It should be noted that the biomass in the liquid phase was primarily composed of bacterial biomass. Some mycelial fungal components could, nevertheless, have also been associated with the bacterial biomass during its estimation. The amount of fungal biomass present in the co-culture was estimated by measuring the dry weight of the fungal pellets. For the attached co-culture, the combined fungal-bacterial biomass and the biomass present in the suspension was measured on a dry weight basis.

3.2.3.2 Phenol and metabolite analysis

The concentration of phenol and volatile fatty acids were monitored by a gas chromatograph (GC) fitted with a flame ionization detector (FID) (Varian BV 430, Middleburg, The Netherlands), according to the procedure described by Hashemi et al. (2015). 1 mL of liquid sample was withdrawn from the batch bottles, filtered and 1 µL of sample was injected to the GC. The samples were diluted 3 times for the measurement of phenol concentrations exceeding 400 mg/L. 50 µL of formic-isovaleric acid was added as an internal standard.

3.2.3.3 Scanning electron microscopy (SEM) - energy dispersive X-ray (EDAX) and transmission electron microscopy (TEM) analysis

A scanning electron microscope (Carl Zeiss, EVO 18, Germany) operating with an accelerating voltage of 15 kV was used to study the distribution of the nanoparticles in the biomass of both the suspended and attached co-culture systems. The samples were dried, fixed with 0.01% of glutaraldehyde solution on carbon coated plates, gold coated using a gold coater equipment (Hitachi, Model E1010, Japan) and subjected to SEM analysis. Later, EDAX was performed on the same samples to see the spatial distribution of the nanoparticles within the fungal-bacterial biomass. Approximately 1 mL of sample was collected after completion of the phenol degradation experiments from the fungal-bacterial suspended co-culture (120 h incubation) and the fungal-bacterial biofilm co-culture systems. About 10 µL of a dilute suspension of the culture medium containing the Se(0) was placed on a carbon coated Cu grid, which was then dried under a table lamp and stained subsequently for 3 min with Ruthenium vapour. A transmission electron microscope (JEOL JEM 2100 HR with EELS), with an accelerating voltage of 200 kV was used to analyze the particle size and shape of the Se(0) in the liquid medium.

3.2.3.4 Dynamic light scattering (DLS) for zeta potential and size distribution

The zeta potential and particle size distribution of the Se(0) produced in the two different co-culture systems were determined by the dynamic light scattering technique (DLS-MALVERN, USA). Under identical conditions of the solution viscosity (0.8872 cp) and dispersant refractive index (1.330) in water as the dispersion medium, samples collected (after 120 h of incubation) from both the suspended and attached co-culture were analysed.

3.2.3.5 Fourier transform - infrared spectroscopy (FT-IR) analysis

The Se(0) and biomass of the co-culture system were finely ground and powdered by a fiber microtome and then blended with KBr. The mixture was converted into ultra-thin pellets by the application of pressure using a hydraulic pressure system (Atlas Manual Hydraulic Press 15T, USA). The liquid samples of the degraded phenol medium were analysed in attenuated total reflection (ATR) mode of the FT-IR (Jasco FT-IR 6300, UK), as described by Mandal and Chakrabarty (2011).

3.3. Results

3.3.1 Suspended growth co-culture incubations

3.3.1.1 Biodegradation of phenol

The co-culture of *P. chrysosporium* and *D. lacustris* completely degraded up to 800 mg/L phenol with the simultaneous reduction of selenite to Se(0) at an intial pH of 6.5 within 120 h (Figure 3.1a). The phenol degradation was almost negligible (7.2%) during the first 48 h of incubation. At concentrations of 1000 and 1025 mg/L of phenol, 987 and 800 mg/L of phenol were degraded, respectively, in 120 h and the residual phenol remained non-degraded. At the highest phenol concentration tested, i.e. 1200 mg/L, the degradation was completely inhibited. The degradation rate varied from 1.33 mg/L/h (for an initial concentration of 100 mg/L) to 12.9 mg/L/h (for an initial concentration of 1025 mg/L). However, it is noteworthy to mention that an active sole culture of *D. lacustris* was unable to degrade phenol even at low concentrations (10 mg/L).

3.3.1.2 Reduction of selenite ions

The suspended growth co-culture reduced about 10 mg/L of selenite ions to Se(0) and other selenium compounds in 72 h, with the simultaneous and complete degradation of 100-400 mg/L phenol (Figure 3.1b). The selenite concentrations decreased from 8 to 0 mg/L in the presence of 600-1100 mg/L phenol, respectively. Selenite was partially reduced to Se(0) by the fungus *P. chrysosporium*, as visualised by the orange-red colour of the originally white coloured fungus. *D. lacustris* was also found to reduce selenite to Se(0) as visualised by the orange-red colour of the bacterial biomass in the suspension. As *D. lacustris* and *P.*

chrysosporium were growing in co-culture, the exact amount of selenite reduced by each of the two microorganisms could not be determined.

3.3.1.3 Acetic acid production

During phenol degradation by the co-culture, varying concentrations of acetic acid were produced in the batch incubations. The highest acetic acid concentration produced was 123 mg/L, in 120 h, by the suspended co-culture system containing an initial phenol concentration of 400 mg/L (Figure 3.1c). No acetic acid production was observed during the first 48 h of co-culture incubation. Acetic acid was produced upon the initiation of the phenol degradation, at concentrations ranging between 100 and 1100 mg/L. As no other carbon source was present in the medium, acetic acid was likely a metabolite of the phenol degradation.

3.3.1.4 Biomass concentration profiles

The biomass concentration in the liquid medium increased to ~0.2 g/L of dry weight after 48 h of incubation (Figure 3.1d). The highest (~1 g/L dry weight) biomass concentration was achieved during the incubation with 600 mg/L of phenol. Thereafter, the final biomass concentration decreased to 0.55, 0.62 and 0.43 g/L in incubations with 800, 1000 and 1025 mg/L of phenol. The biomass consisted mainly of bacterial biomass, but still some fungal filaments were present. There was no significant change in the biomass of the fungal pellets until 48 h; however, afterwards, the fungal pellets began to lose some of the hairy filaments at phenol concentrations of 400 and 600 mg/L (Figure 3.1e).

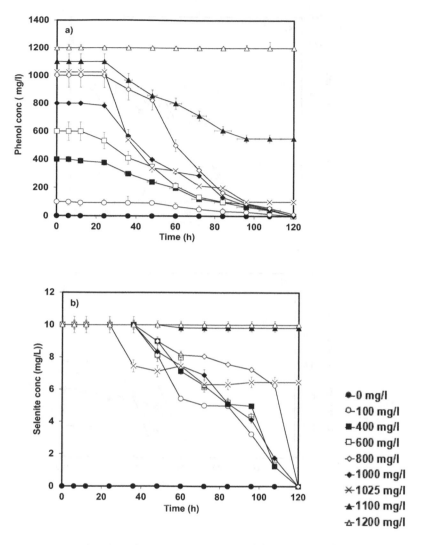

Figure 3.1 Phenol degradation by suspended co-culture of P. chrysosporium and D. lacustris: (a) degradation of phenol, (b) reduction of selenite,

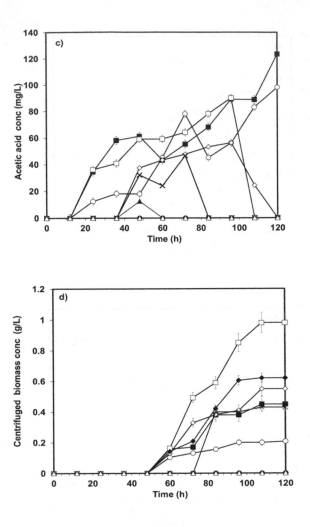

Figure 3.1 Phenol degradation by suspended co-culture of P. chrysosporium and D. lacustris: (c) production of acetic acid, (d) profile of centrifuged biomass

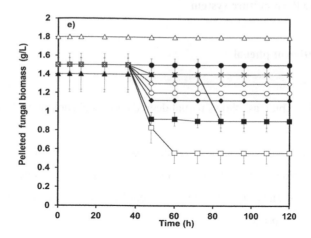

Figure 3.1 Phenol degradation by suspended co-culture of P. chrysosporium and D. lacustris: (e) profile of the growth of pelleted fungal biomass

3.3.2 Attached growth co-culture system

3.3.2.1 Biodegradation of phenol

The attached co-culture biomass shows partial degradation of phenol, up to 100 mg/L (Figure 2a) within 120 h (at the rate of 0.83 mg/L/h), whereas phenol degradation was completely inhibited at only 200 mg/L. The degradation decreased probably due to the presence of the layer of the bacterial biomass over the fungal biomass that presumably decreased the contact of the fungus to phenol. Thus, an increase in the phenol concentrations from 100 to 200 mg/L inhibited the degradation process.

3.3.2.2 Reduction of selenite ions

The attached fungal-bacterial biomass culture was found to reduce up to 6.88 mg/L of selenite ions within 72 h (Figure 3.2b). The Se(0) was found to be sequestered by the fungal biomass of the attached growth co-culture system and the colour was orange-red. In contrast to suspended growth co-culture, no orange-red coloration was observed in the bacterial biomass attached to the fungus. Concerning biomass growth, there was no significant increase in the biomass content in the liquid medium and the fungal-bacterial pellets.

3.3.2.3 Acetic acid production

Acetic acid was still produced despite the much lower amount of phenol degraded (Figure 3.2c). At an initial concentration of 10 mg/L phenol, no acetic acid was produced. However, at an initial phenol concentration of 50 mg/L, 33 mg/L of acetic acid was produced after 60 h of incubation of the co-culture. For an initial phenol concentration of 100 mg/L, the acetic acid concentration amounted to 56 mg/L after 60 h incubation, after that, the acetic acid concentration was almost constant.

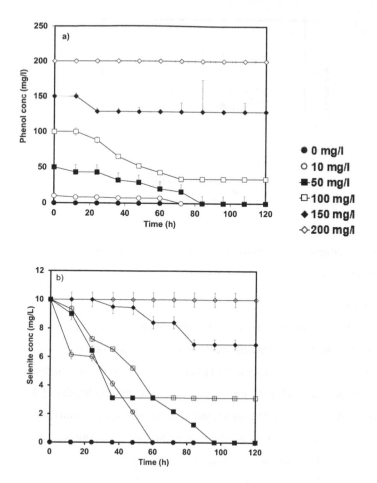

Figure 3.2 Phenol degradation by attached co-culture of P. chrysosporium and D. lacustris:
(a) degradation of phenol (b) reduction of selenite

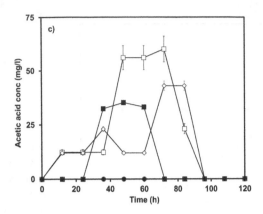

Figure 3.2 Phenol degradation by attached co-culture of P. chrysosporium and D. lacustris (c) production of acetic acid.

3.3.3 FT-IR analysis

The phenol degradation was evident from the disappearance and appearance of related functional groups. Figure 3a shows the IR spectrum of the initial co-culture system containing mostly phenol. The spectrum resembles that of phenol with the following characteristic peaks: (i) a broad band in the region of 3489 cm^{-1} representing the intermolecular H–bonded O–H. The O–H stretching of water and its intermolecular H bonding overlaps in this region (observed only if water is present in the system), and ii) the aromatic ring stretching of C–H, indicated by the small peak at 3045 cm^{-1}, whereas the C ring stretching is given by the presence of multiple peaks at 1596.43, 1499.97 and 1474 cm^{-1}, respectively. The OH bonding is represented by the peak at 1368.23 cm^{-1}, while the C–O stretching is given by 1234.66 cm^{-1}. The peaks at 811.18 cm^{-1} and 753.66 cm^{-1} represented the C–H bending, respectively.

The FT-IR spectra of the co-culture system of the liquid medium after biodegradation were different from the initial co-culture medium. The aromatic C–H stretching as observed previously at 3054 cm^{-1} was not present in the co-culture system of the liquid medium. Besides, C=C ring stretching at 1495 cm^{-1} and 1468 cm^{-1}, which usually appear for phenol, was also not present. C–O stretching for phenol observed at 1220 cm^{-1} was absent. In-plane O–H bending of phenol at 1360 cm^{-1} does not appear in the system. On the other hand, a new peak appeared in the region of 1725-1720 cm^{-1}, indicating the stretching due to the C=O bond in acids (e.g. cis-cis muconic acids). The sharp peak in the region of 2305 cm^{-1} indicates the presence of C=C bonds of non-aromatic compounds which might be an intermediate compound that formed during phenol degradation. The large peak appearing in the region 3400-3300 cm^{-1} might be indicating hydrogen bound O-H stretching. This peak of phenol overlaps with the O–H peak of water present in the medium. These shifts in peaks confirm that phenol degradation has occurred in the incubations.

The FT-IR spectra of the co-culture also shows that thiols, i.e. S-H groups (Figure 3b) are present, typically represented by the peak at 2850 cm^{-1} (Banwell and McCash, 1983). The S-H group may be related to glutathione and the glutathione peroxidase enzyme helps in the detoxification of selenite ions to Se(0) according to Eq. (1) (Kessi and Hanselmann 2004):

$$S-H- (Glutathione) + Selenite \rightarrow Se(0) \qquad (1)$$

The FT-IR spectra of the Se(0) produced as a result of the suspended co-culture system contains a sharp peak at 2066 cm^{-1} (Figure 3c), indicating the presence of carbonyl complexes attached to the nanoparticles. A small band or peak near 2066 cm^{-1} was observed and based on the information available in the literature (Vessieres et al. 1999), this indicates that some carbonyl complexes of various transition elements are present in this region. In the present case, with only selenite present in the incubation system, Se can form a complex with the carboxylic compounds derived from the phenol degradation.

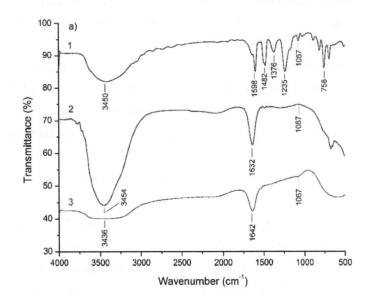

Figure 3.3 (a) FT-IR spectrum of phenol degradation by suspended co-culture. Aromatic O-H bond 3450 cm⁻¹, 3454 cm⁻¹ and 3436 cm⁻¹; Aromatic C=C bond at 1598 cm⁻¹ and 1482 cm⁻¹; Amide C=O stretch at 1632 cm⁻¹ and 1642 cm⁻¹; Acidic C-O bond at 1235 cm⁻¹; Ether/Alcohol C-O bond at 1067 cm⁻¹; 1. Initial medium with phenol at 800 mg/L, 2. Intermediate after 60 h, 3. Absence of aromatic C=C bond shows completely degraded medium after 120 h in suspended co-culture medium

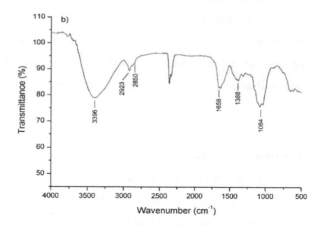

Figure 3.3 (b) FT-IR of biomass: O-H group at 3435 cm^{-1}, S-H group at 2066 cm^{-1}, C-O-N-H$_2$ of polypeptide at 1638 cm^{-1}

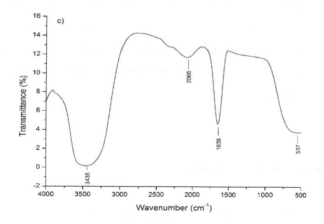

Figure 3.3 (c) FT-IR spectrum of the Se(0) produced in suspended co-culture medium: OH group at 3461.72 cm^{-1}, C-O complex at 2066 cm^{-1}, proteins and peptide N-H at 1636.62 cm^{-1}, respectively

3.3.4. Zeta potential analysis

The suspended fungal-bacterial co-culture had a zeta potential of -28.7 mV (Figure 3.S1a), while the attached fungal bacterial co-culture had a zeta potential of -12.9 mV (Figure 3.S1b), respectively. The conductivity of the suspended co-culture system was 0.207 mS/cm. In the suspended co-culture system, all the selenium particles had nanometric dimensions, which was interpreted from the curve size of the plot between diameter and intensity (%), with the highest at ~75 nm. A small width of the distribution curve indicates uniformity of the particle size, i.e. the lower end fractions are on an average close to the higher end fragments in the statistical distribution of particles (Figure 3.3d). This was further supported by the polydispersity index, which was found to be less than unity. In attached co-culture system, Se(0) had an average particle size of ~250 nm in diameter. Nevertheless, the visible particles has sizes >100 nm (Figure 3.3d.1).

3.3.5. SEM-EDAX analysis of the biomass

The SEM analysis (Figs. 3.4a and 3.4b) reveals mainly changes in the structural morphology of the fungus predominating the fungal-bacterial co-culture. The suspended co-culture exhibited a spongy biomass and has a non-layered structure. In contrast, the attached growth system exhibited a layered structure where the reduced Se(0) particles spread in a non-uniform manner over each of the layers. This network of cells is quite distinct, in which the bacterial cells may be embedded. The cells appear to have more depth compared to those of the attached growth system. The suspended co-culture also shows a non-uniform distribution of the reduced selenium, which was present not only on the surface but also deposited on the cell walls. EDAX analysis showed much less sequestration of Se(0) in the visible predominant fungal biomass (Figures 4.4c and 4.4d)

3.3.6 TEM analysis of the liquid phase

The results from TEM analysis after the biodegradation tests reveals the fact that the Se(0) were formed mostly as nanospheres in the liquid medium. Table 2 shows the size distribution of Se(0) produced. For the suspended co-culture system, the Se(0) produced had the smallest diameter of 3.58 nm (Figure 4.4g), while in the attached co-culture system, the smallest size of the Se(0) produced was 58.52 nm (Figure 4.4h). The observed Se(0) particles were mostly nanospheres. The Se(0) produced in the medium was more likely expected be produced by *D. lacustris*, because the Se(0) produced by the fungus is usually entrapped in the fungal biomass (Espinosa-Ortiz et al. 2015).

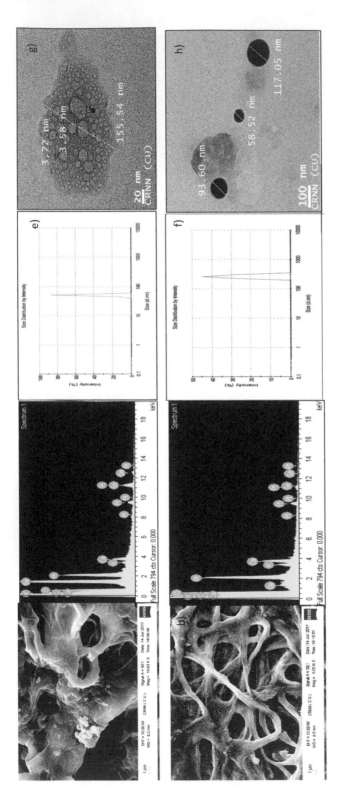

Figure 3.4: Electron microscope analysis of the fungal–bacterial biomass: (a) suspended and (b) attached growth co-culture. EDAX of fungal and fungal-bacterial biomass: (c) suspended and (d) attached co-culture, size distribution of Se(0) in (e) suspended and (f) attached co-culture. TEM of Se(0) produced in (g) suspended and (h) attached co-culture system.

3.4 Discussion

3.4.1 Phenol removal by suspended and attached co-cultures of *P. chrysosporium* and *D. lacustris*

This study showed that the initial pH played an important role in the development of the fungal bacterial co-culture and subsequent degradation of phenol. The pH of 6.5 was found to be suitable for sustaining the metabolism of both the bacterium and the fungus. The initial lag period of 48 h during the uptake of phenol may be attributed to the acclimatization period of the fungal-bacterial system to the medium conditions and viability of the bacterial cells only increased with the adaptation of the fungus to the incubation conditions. However, *P. chrysosporium* was observed to degrade up to 400 mg/L of phenol at an initial pH of 4.5 (Werkeneh et al. 2017). The suspended co-culture of *P. chrysosporium* and *D. lacustris* degraded phenol at a concentration of 1025 mg/L at a maximum rate of 12.9 mg/L/h. Attached co-culture degraded phenol at an initial concentration of 100 mg/L at a maximum rate of 0.83 mg/L/h, but phenol degradation is completely inhibited at 200 mg/L. This decrease may be due to the presence of bacterial biomass on the fungus On the other hand, *D. lacustris* could not grow on phenol and reduce selenite simultaneously at pH ranging from 5.0-7.0 (data not shown). Therefore, the bacterium grew on the intermediates produced by the fungus, e.g. aromatic acid like cis-cis muconic acid, detected from FT-IR analysis (Figure 4.3a).

3.4.2 Analysis of the mechanism of selenite reduction

The reduction of selenite to Se(0) followed the same pattern as the phenol degradation (Figures 4.1a and 1b), emphasizing the simultaneous detoxification of selenite and phenol. The FT-IR analysis of the biomass reveals the generation of S-H groups bound to glutathione that facilitated the detoxification of selenite ions (Kessi and Hanselmann 2004). The Se(0) appeared in the medium only after 48 h of incubation indicating the fact that, after acclimatization and initiation of the degradation of phenol, the intermediates of phenol degradation were utilised by the bacterium for the reduction of selenite to Se(0). After 60 h, however, the fungal pellets turned orange-red in colour (Figure 6), indicating the bioconversion of selenite to Se(0).

The selenite reduction by the attached-growth incubation was more efficient compared to the suspended growth culture, i.e. 6.88 mg/L of selenite was degraded in the presence of 150 mg/L phenol. On the contrary, the suspended co-culture incubation was more efficient for the degradation of phenol. In the latter system, the layer of the bacterium surrounding the fungal pellets hindered the availability of oxygen and growth of the fungus, its metabolism, and subsequently, the uptake of phenol and selenite.

3.4.3 Proposed mechanism for the degradation of phenol coupled to selenite reduction

The fungus likely initiated phenol degradation with the production of an intermediate metabolite, which was consumed by the bacterium (Figure 4.5). The presence of the metabolite stimulated phenol degradation and increased the degradation capacity of the fungus. Production of acetic acid upon initiation of the phenol degradation and the depletion of acetic acid later, indicates its consumption. Hence, it is likely an intermediate metabolite utilised by the bacterium. The bacterial layer outside the fungal structure in the attached co-culture system most likely reduces the availability of phenol to the fungus, resulting in decreased degradation and hence the reduced concentration of the metabolite. Thus, the bacterial biofilm poses a major mass transfer limitation to the substrate uptake. Moreover, the respiratory activity of *P. chrysosporium* can also be negatively affected or inhibited by the multi-layered bacterial growth over the fungus, the presence of selenite ions and anoxic zones, which has also been reported in *P. chrysosporium* growing in drip-flow bioreactor configuration (Espinosa-Ortiz et al. 2016).

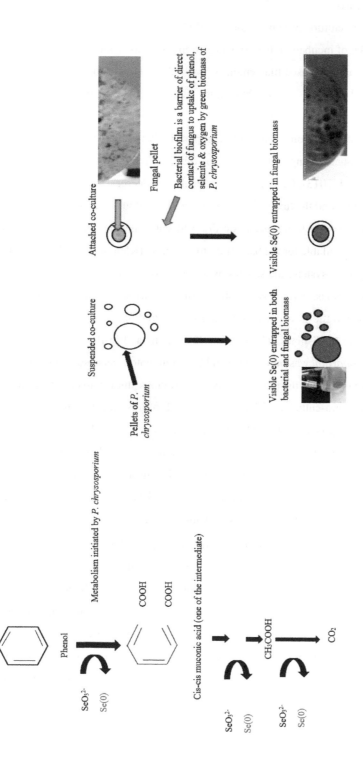

Figure 3.5: Proposed mechanism of phenol degradation with simultaneous reduction of selenite by the fungal-bacterial co-culture of P. chrysosporium and D. lacustris with a schematic representation showing the role suspended and co-cultures of P. chrysosporium and D. lacustris during the biodegradation of phenol coupled to selenite reduction in batch incubations.

3.4.4 Biomass growth

In the suspended co-culture system, although a slight reduction in the biomass content was observed after 48 h of incubation, the dry weight remained constant till 120 h (Figure 3.1e) indicating spongy and increased filamentous growth. After 48 h of incubation, the bacterial biomass was found to increase in weight (Figure 3.1d). The delayed growth of the bacterial cells can be attributed to the adaptation of the bacterial cells to the phenol containing medium and the bioconversion of phenol to comparatively more bioavailable and less toxic intermediate compounds. Fungal biomass has also been reported to adsorb phenol from the medium (Farkas et al. 2013). Pelletization of fungal cells decreases with an increase in the phenol concentrations and the fungus becomes more filamentous (Figure 3.3a and 3.5). This filamentous growth is due to its metabolic drive to consume the nutrients and make the necessary nutrients available for the bacterium to consume (Klein and Paschke 2004).

In the attached growth system, the selenite was found to be mostly reduced and localized inside the fungal biomass with a white sheath of the bacterium surrounding the fungal biomass. Compared to the suspended co-culture system, red Se(0) was less prevalent in the medium. According to Espinosa-Ortiz et al. (2016), in a drip-flow bioreactor treating selenite, *P. chrysosporium* was more dense and less filamentous as observed in the attached biofilm co-culture system of the present study. All previous studies involving selenite have only focused on mono-cultures of fungal or yeast biofilms like *Candida albicans* and *Candida topicalis* (Espinosa-Ortiz et al. 2016; Harrison et al. 2007; Werkeneh et al. 2017). These studies have reported that the presence of selenite had inhibited hyphae formation, contamination with other bacteria and change in the biofilm structure and composition after prolonged treatment. Hence, further morphological studies are required to ascertain the effect of selenite on the co-culture composition and growth pattern.

3.4.5 Zeta potential and size distribution of the Se(0) particles

The suspended growth co-culture system had a zeta potential of -28.7 mV and an average particle size of 70 nm (Figure 4.4e). In the facultative anaerobe *Citrobacter freundii* Y9, the average zeta potential value was -20.61 (± 1.29) mV for EPS (extra polymeric substance) in groundwater and -28.02 (± 0.08) mV for EPS in soil solution, respectively. These negative values of zeta potential have shown to stabilise the Se(0) particles (Wang et al. 2018). On comparing the two different co-culture systems, the suspended co-culture system was found to produce more stable Se(0) particles as evidenced from the relatively lower negative zeta potential of -28.7 mV in relation to -12.9 mV observed with the attached fungal-bacterial co-culture.

For the suspended co-culture system, the Se(0) produced had the smallest diameter of 3.58 nm (Figure 4.4g). In this system, there could be a layer of EPS surrounding the nanoparticles, which possibly provided the necessary zeta potential and hence the stability. However, this was not so much pronounced in the attached co-culture system and hence the Se(0) particles exhibited a tendency to coalesce and form aggregates. This also led to the formation of relatively larger sized Se(0) particles in the attached co-culture system, wherein the smallest size of the Se(0) produced was 58.52 nm (Figure 4.4h). In order to elucidate the factors that govern the size distribution of the produced Se(0) particles in co-culture systems, more in-depth studies are required to characterise the nature of EPS produced during the simultaneous removal of phenol and selenite.

3.5 Practical applications

P. chrysosporium is an efficient polyaromatic hydrocarbon (PAH) degrading white-rot fungus (Huang et al. 2017; Paszczynski and Crawford 1995). In a recent study, the fungus improved the degradation capacity of an activated sludge reactor treating coking wastewater (Haieli et al. 2017). In another study involving mixed biocatalysts, the fungus reduced the toxicity of Pb^{2+} to the native microorganisms of a Pb-contaminated agricultural waste composting site. The fungus is also a selenite reducing organism, capable of degrading phenol (400 mg/L) in the presence of selenite ions (15 mg/L) (Espinosa-Ortiz et al. 2015; Werkeneh et al. 2017). In a continuously operated up flow bioreactor inoculated with *P. chrysosporium*, removal efficiencies ranging from 50-75% (at a critical loading rate of 12 mg/L/h) and 75-90.8% (at a critical loading rate of 4.3 mg/L/h) have been achieved (Werkeneh et al. 2017).

The present study focussed on the capability of *P. chrysosporium* to stimulate the activity of another Se-reducing bacterium, i.e. *D. lacustris*, which is not capable of degrading phenol (Figure 1d). Refinery effluents have a neutral to alkaline pH (7.0-8.0), where the co-culture of *P. chrysosporium* and *D. lacustris* can sustain, whereas the single culture of *P. chrysosporium* was only able to degrade 67.3% of 400 mg/L phenol at an acidic pH of 4.5. Moreover, solo culture of *P. chrysosporium* gets completely inhibited at phenol concentrations of 600 mg/L. About 800 mg/L of phenol was completely degraded in batch incubations of the suspended co-culture and the inhibitory concentration of phenol was 1200 mg/L. Thus, co-metabolic degradation of phenol is more feasible at higher pH (7-8), and the efficiency of phenol degradation is almost doubled. Additionally, the reduction of selenite in the co-culture system produces Se(0) nanoparticles, which has potential anticancer effect, antimicrobial activity and has several applications in the field of pharmaceutical, medical and environmental sciences (Guleria et al. 2018). PAH containing wastewater also contains toxic metal ions like Cr^{2+}, Cd^{2+}, Zn^{2+}, As^{3+}, As^{5+}, Hg^{2+} (Liu et al. 2017) which may affect the selenite reducing capacity of the co-culture. More experiments with different concentrations and combinations of heavy metals are required to be performed to find the exact effects of these metal ions and the effect of toxicity.

The capacity of the suspended growth co-culture system to degrade phenol was found to be higher compared to the attached growth co-culture system. Surprisingly, the two different co-culture methods produced Se(0) of different size ranges by reducing selenite in the presence of phenol as electron donor. In the suspended growth co-culture system, a filamentous and sponge-like fungal morphology was apparent, whereas a rigid and layered fungal-bacterial morphology was conspicuous in the attached growth co-culture system

3.6 Conclusions

Batch experiments reveal 800 mg/L of phenol was completely degraded in batch incubations of the suspended co-culture of *Phanerochaete chrysosporium* and *Delftia lacustris* and the inhibitory concentration of phenol was 1200 mg/L. Attached co-culture of *Phanerochaete chrysosporium* and *Delftia lacustris* was able to degrade completely 50 mg/L phenol, but was completely inhibited at a phenol concentration of 200 mg/L. The selenite ions were biologically reduced to extracellular Se(0) nanoparticles of 3.58 nm diameter and 58.5 nm diameter by suspended co-culture.

References

Abdelwahab, O., Amin, N.K., El-Ashtoukhy, E.Z. Electrochemical removal of   phenol from oil refinery wastewater. J. Hazard. Mater. 163 (2009):711-716.

Almendariz, F.J., Meraz, M., Olmos, A.D., Monroy, O. Phenolic refinery wastewater biodegradation by expanded granular sludge bed bioreactor. Water. Sci. Technol. 52 (2005): 391-396.

Banwell, C.N., McCash, E.M. Fundamentals of Molecular Spectroscopy. McGraw Hill Education, Europe. 1983.

Cheng, Z., Li, C., Kennes, C., Ye, J., Chen, D., Zhang, S., Chen J., Yu, J. Improved biodegradation potential of chlorobenzene by a mixed fungal-bacterial consortium. Int. J. Biodet. Biodeg. 123 (2017): 276-285.

Espinosa-Ortiz, E.J., Gonzalez-Gil, G., Saikaly, P.E., Hellebusch, EDv., Lens, P.N.L. Effects of selenium oxyanions on the white-rot fungus *Phanerochaete chrysosporium*. Appl. Microbiol. Biotechnol. 99 (2015): 2405-2418.

Espinosa-Ortiz, E.J., Pechaud, Y., Lauchnor, E., Rene, E.R., Gerlach, R., Peyton, B.M., Lens, P.N.L. Effect of selenite on the morphology and respiratory activity of *Phanerochaete chrysosporium* biofilms. Bioresour. Technol. 210 (2016): 138-145.

Farkas, V., Felinger, A., Hegedűsova, A., Dékány, I., Pernyeszi, T. Comparative study of the kinetics and equilibrium of phenol biosorption on immobilized white-rot fungus *Phanerochaete chrysosporium* from aqueous solution. Coll. Surf. B: Bioint. 103 (2013): 381-390.

Gadd, G.M. Geomycol: biogeochemical transformations of rocks, minerals, metals and radionuclides by fungi, bio weathering and bioremediation. Mycol. Res. 11 (2007): 3-49.

Guleria, A., Chakraborty, S., Neogy, S., Maurya, D.K., Adhikari, S. Controlling the phase and morphology of amorphous Se nanoparticles: their prolonged stabilization and anticancer efficacy. Chem. Comm. 54 (2018): 8753-8756.

Hailei, W., Ping, L., Ying, W., Lei, L., Jianming, Y. Metagenomic insight into the bioaugmentation mechanism of *Phanerochaete chrysosporium* in an activated sludge system treating coking wastewater. J. Hazard. Mater. 32 (2017): 820-829.

Harrison, J.J., Ceri, H., Yerly, J., Rabiei, M., Hu, Y., Marinuzzi, R., Turner, R.J. Metal ions may suppressor enhance cellular differentiation in *Candida albicans* and *Candida tropicalis* biofilms. Appl. Environ. Microbiol. 73 (2007): 4940-4949.

Hashemi, W.l., Maraqa, M.A., Rao, M.V., Hossain, M.M. Characterization and removal of phenolic compounds from condensate-oil refinery wastewater. Desal. Water. Treat. 54 (2015): 660- 671.

Huang, C., Zeng, G., Huang, D., Lai, C., Xu, P., Zhang, C., Cheng, M., Wan, J., Liang, H., Zhang, Y. Effect of *Phanerochaete chrysosporium* inoculation on bacterial community and metal stabilization in lead-contaminated agricultural waste composting. Bioresour. Technol. 243 (2017): 294-303.

Jørgensen, N.O., Brandt, K.K., Nybroe, O., Hansen, M. *Delftia lacustris* sp. nov, a peptidoglycan-degrading bacterium from fresh water, and emended description of *Delftia tsuruhatensis* as a peptidoglycan-degrading bacterium. Int. J. Syst. Evol. Microbiol. 59 (2009): 2195- 2199.

Kessi, J., Hanselmann, K.W. Similarities between the abiotic reductions of selenite with glutathione and the dissimilatory reaction mediated by *Rhodospirillum rubrum* and *Escherichia coli*. J. Biolog. Chem. 279 (2004): 50662-50669.

Klein, D.A., Paschke, M.W. Filamentous fungi: the indeterminate lifestyle and microbial ecology. Microb. Eco. 47 (2004): 224-235.

Lawson, S., Macy, J.M. Bioremediation of selenite in oil refinery wastewater. Appl. Microbiol. Biotechnol. 43 (1995): 762-765.

Li, D.B., Cheng, Y.Y., Wu, C., Li, W.W., Li, N., Yang, Z.C., Tong, Z.H., Yu, H.Q. Selenite reduction by *Shewanella oneidensis* MR-1 is mediated by fumarate reductase in periplasm. Sci. Rep. 4 (2014): 1-7.

Liu, S.H., Zeng, G.M., Niu, Q.Y., Liu, Y., Zhou, L., Jiang, L.H., Tan, X.F., Xu, P., Zhang, C. Cheng, M., 2017. Bioremediation mechanisms of combined pollution of PAHs and heavy metals by bacteria and fungi: A mini review. Bioresour. Technol. 224(2017): 25-33.

Mandal, A., Chakrabarty, D. Isolation of nanocellulose from waste sugarcane bagasse (SCB) and its characterization. Carbohyd. Pol. 86(2011): 1291-1299.

Nurdogan, Y., Evans, P., Sonstegard, J. Selenium removal from oil refinery wastewater using advanced biological metal removal (ABMet®) process. In: Proceedings of the Water Environment Federation, WEFTEC, Session 1-10 (2012), 229-241.

Paszczynski, A., Crawford, R.L. Potential for bioremediation of xenobiotic compounds by the white-rot fungus *Phanerochaete chrysosporium*. Biotechnol. Prog. 11 (1995): 368-379.

Pradeep, N.V., Anupama, S., Navya, K., Shalini, H.N., Idris, M., Hampannavar, U.S. Biological removal of phenol from wastewaters: a mini review. Appl. Water. Sci. 5 (2015), 105-112.

Presentato, A., Piacenza, E., Anikovskiy, M., Cappelletti, M., Zannoni, D., Turner, R.J. Biosynthesis of selenium nanoparticles and nanorods as a product of selenite bioconversion by the aerobic bacterium *Rhodococcus aetherivorans* BCP1. New Biotechnol. 41 (2017), 1-8.

Sarkar, J., Dey, P., Saha, S., Acharya, K. Mycosynthesis of selenium nanoparticles. IET. Micro. Nano. Lett. 6 (2011): 599-602.

Tam, K., Ho, C.T., Lee, J.H., Lai, M., Chang, C.H., Rheem, Y., Chen, W., Hur, H.G., Myung, N.V. Growth mechanism of amorphous selenium nanoparticles synthesized by *Shewanella sp*. HN-41. Biosci. Biotechnol. Biochem. 74 (2010): 696-700.

Tan, L.C., Nancharaiah, Y.V., Hullebusch, EDv., Lens, P.N.L. Selenium: environmental significance, pollution, and biological treatment technologies. Biotechnol. Adv. 34 (2016): 886-907.

Vessieres, A., Salmain, M., Brossier, P., Jaouen, G. Carbonyl metallo immuno assay: a new application for Fourier transform infrared spectroscopy. J. Pharma. Biomed. Anal. 21 (1999): 625-633.

Wadgaonkar, S.L., Nachariah, V.Y., Jacob, C., Esposito, G., Lens, P.N.L. Selenate reduction by *Delftia lacustris* under aerobic condition. J. Microbiol. (2019) (In Press).

Wang, X., Song, W., Qian, H., Zhang, D., Pan, X., and Gadd, G.M. Stabilizing interaction of exopolymers with nano Se and impact on mercury immobilization in soil and groundwater. Env. Sci. Nano. 5(2018): 456-466.

Werkeneh, A.A., Rene, E.R., Lens, P.N.L. Simultaneous removal of selenite and phenol from wastewater in an up flow fungal pellet bioreactor. J. Chem. Technol. Biotechnol. 93 (2017): 1003-1011.

Supplementary materials

Results

	Mean (mV):	Area (%)	Width (mV):
Zeta Potential (mV): -28.7	Peak 1: -25.3	39.9	4.83
Zeta SD (mV): 13.2	Peak 2: -40.9	39.4	6.61
Conductivity (mS/cm): 0.207	Peak 3: -10.6	20.7	4.56

Zeta potential out of range

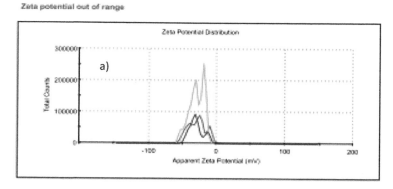

Results

	Mean (mV):	Area (%)	Width (mV):
Zeta Potential (mV): -12.9	Peak 1: -12.9	100.0	5.89
Zeta SD (mV): 5.89	Peak 2: 0.00	0.0	0.00
Conductivity (mS/cm): 0.119	Peak 3: 0.00	0.0	0.00

Zeta potential out of range   Conductivity is out of range - check cell or sample

Figure 3.S1. Zeta potential values and particle size distribution of: (a) suspended and (b) attached co-culture system

Chapter 4

Bioconversion of gaseous effluents of oil-refinery by production of acids and alcohols from CO

This chapter has been published as:

Chakraborty, S., Rene, E.R., Lens, P.N., Veiga, M.C., Kennes, C., 2019. Enrichment of a solventogenic anaerobic sludge converting carbon monoxide and syngas into acids and alcohols. *Bioresource. Technology*. 272: 130-136.

Abstract

An anaerobic granular sludge was acclimatized to utilize CO in a continuously gas-fed stirred tank bioreactor by applying operating conditions expected to stimulate solventogenesis, i.e. the production of alcohols, and allowing to enrich for solventogenic populations. A cycle of high (6.2) and low (4.9) pH was applied in order to produce volatile fatty acids first at high pH, followed by their bioconversion into alcohols at low pH. The addition of yeast extract stimulated biomass growth, but not necessarily solventogenesis. The highest concentrations of metabolites achieved were 6.18 g/L acetic acid (30th day), 1.18 g/L butyric acid (28th day), and 0.423 g/L hexanoic acid (32nd day). Subsequently, acids were metabolized at lower pH, producing alcohols at concentrations of 11.1 g/L ethanol (43rd day), 1.8 g/L butanol (41st day) and 1.46 g/L hexanol (42nd day), confirming the successful enrichment strategy. Similarly, the enriched sludge could also convert syngas into acids and alcohols.

Keywords: ethanol; butanol; hexanol; volatile fatty acids; solventogenesis

4.1 Introduction

Carbon monoxide (CO) is a major compound found in waste gases from some industries such as steel producing plants and it is also a major component of synthesis gas or syngas (Abubackar et al., 2011). In biorefineries, CO-rich syngas can be obtained from the gasification of biomass, solid waste or other carbonaceous feedstocks. Some of those feedstocks, such as lignocellulosic biomass, can also be hydrolyzed to obtain carbohydrates that can then similarly be fermented to generate high value metabolites, biofuels or platform chemicals (Fillat et al., 2017).

CO can be converted anaerobically by acetogenic bacteria into a spectrum of commercially useful metabolites. This gas-fermentation technology represents an environmentally-friendly alternative to the more conventional crude oil-based refineries or other petrochemical processes for the production of fuels or chemicals. Several acetogenic bacteria assimilating CO are also able to metabolize CO_2 and H_2 as well as syngas, e.g. *Clostridium ljungdahlii, Clostridium autoethanogenum, Clostridium ragsdalei, Clostridium carboxidivorans* (Fernández-Naveira et al., 2016 a; Shen et al., 2014, Huhnke et al., 2010, Philips et al., 2015). Recent research on CO, CO_2/H_2 and syngas fermentation has focused on the production of different end-products, such as ethanol (Groenestijn et al., 2013), higher alcohols such as butanol or hexanol (Fernández-Naveira et al., 2017a, Philips et al., 2015), 2, 3-butanediol (Köpke et al., 2011), biomethane (Sancho et al., 2016) or biopolymers, e.g., polyhydroxyalkanoates (Lagoa-Costa et al., 2017). Some biofuels obtained from C1-gas fermentation like biogas or biomethane (López et al., 2012) may need to be upgraded to remove CO_2 or other by-products such as hydrogen sulphide (López et al., 2013). Many studies have been done with available pure cultures and only a few with mixed cultures (Rachbauer et al., 2017; Xu et al., 2015). Metabolic engineering is used as an additional opportunity to broaden the range of potential bio commodities that can be produced through anaerobic gas (CO, syngas and waste gas effluents) fermentation (Bengelsdorf et al., 2013). Bio alcohols such as ethanol and butanol, among others, are interesting (bio)fuels that can be obtained through anaerobic gas fermentation and are suitable to replace gasoline or to be mixed with it in different ratios (Fernández-Naveira et al., 2017a). So far, only one pure acetogenic strain has unequivocally been described to produce higher alcohols through such gas fermentation, namely *C. carboxidivorans* (Fernández-Naveira et al., 2017 a, b, Philips et al., 2015). Other bacteria should potentially be able to perform similar metabolic conversions. Acetogens generally first convert gases into organic acids at relatively high (near-neutral) or slightly acidic pH, e.g. pH 6. The production of these acids induces a pH drop if the latter is uncontrolled. Such acidic conditions will then stimulate solventogenesis and the production of alcohols, i.e. either ethanol or higher alcohols.

Although some studies have been done on the enrichment of ethanol producing communities (Singla et al., 2014), to the best of our knowledge no thorough enrichment studies have been performed for the production of higher alcohols through anaerobic gas fermentation. A recent study on the enrichment of a mixed culture has focused on the production of acetate (350 mg/l) in batch bottles by BES (50 mM) treated anaerobic sludge in the presence of biogas containing CO_2, H_2 and CH_4 (Omar et al., 2018).—The medium in that study was supplemented with vitamins. Besides, a hollow fibre membrane biofilm reactors has been found to produce acetate (4.22 g/L), butyrate (1.35 g/L), caproate (0.88 g/L) and caprylate (0.52 g/l) at 35°C, but no alcohols were detected (Shen et al., 2018). Modulating the partial pressure composition, i.e. p H_2 and pCO_2 (Wang et al., 2018), the amount of alcohol produced by a mixed culture was 16.9 g/L in a hollow fibre membrane bioreactor.

A mixed culture predominantly composed of *Alkalibacterium bacchi* CP15 and *Clostridium propionicum* enriched from a monoculture syngas fermenting reactor produced, respectively, 1.3 g/L and 2.36 g/L ethanol in medium supplemented with yeast extract and corn steep liquor. According to Liu e al. (2014a), the efficiency of CO utilisation was 40.4 % and 50.3% in medium supplemented with, respectively, YE and corn steep liquor. In a continuous syngas fermenting reactor with cell recycle, a mixed culture of *Alkalibacterium bacchi* CP15 and *Clostridium propionicum* formed a maximum of 8 g/L ethanol, 6 g/L propanol and 1 g/L butanol, corresponding to 61% CO consumption in YE free medium (Liu e al., 2014b).

Therefore, this study aimed to enrich an anaerobic sludge for solventogenic bacteria able to ferment C1 gases, i.e. CO, CO_2 and syngas, to ethanol and higher alcohols, applying specific conditions such as the alternation of high and low pH values of the bioreactor medium or the addition of a specific inhibitor of methanogens (i.e., bromoethane sulphonate, BES) in order to select and facilitate the growth of solventogens instead of methanogens. The effect of yeast extract and L-cysteine-HCl on the metabolism of the solventogenic acetogens was also studied.

4.2. Material and Methods

4.2.1 Biomass and medium composition

The anaerobic sludge (120 mL) was obtained from Biothane Systems International BV (Delft, The Netherlands) and it originated from an anaerobic digester producing biogas from industrial wastewater (Van Lier et al., 2015). The sludge was pregrown on CO (in triplicates in 40 ml bottle) for 14 days at uncontrolled pH (initial pH of 5.5 to favour solventogenesis) at a shaking speed of 120 rpm. The acclimatized sludge of 120 ml was used as inoculum in the reactor. A modified basal anaerobic unautoclaved medium was used as described (NaCl, 0.9 g; $MgCl_2 \cdot 6H_2O$, 0.4 g; KH_2PO_4, 0.75 g; K_2HPO_4 0.5 g; $FeCl_3 \cdot 6H_2O$, 0.0025 g; 0.1% resazurin, 0.75 g) elsewhere (Fernández-Naveira et al., 2016a) with a variation in the initial concentrations of yeast extract (YE - 0.3 g/L) and L-cysteine-HCl (0.6 g/L). No vitamin solution was added to the bioreactor, contrary to other studies focusing on solventogenic clostridia (Fernández-Naveira et al., 2017b, c) as some CO fermenting cultures have been shown to be able to metabolize gases in the absence of vitamins (Abubackar et al., 2015; 2016).

4.2.2 Set-up and operation of the continuous gas-fed bioreactor

Bioreactor experiments were performed in a 2 L BIOFLO 110 bioreactor (New Brunswick Scientific, Edison, NJ, USA) containing 1.2 L medium (working volume). A microsparger was introduced inside the reactor for sparging CO (100 %) as the sole gaseous substrate, fed continuously at a rate of 10 mL/min using a mass flow controller (Aalborg GFC 17, Müllheim, Germany). The temperature of the bioreactor was maintained at 33°C by means of a heating water jacket. Four baffles were symmetrically aligned to avoid vortex formation of the liquid medium and to enhance mixing. The bioreactor was supplied with a DO probe connected to a bio-flow controller and the dissolved oxygen in the bioreactor was less than 0.5% (Fernández-Naveira et al., 2017b, c, d). The agitation speed maintained thoroughout the experiment was 120 rpm. The bioreactor was operated at a constant pH of 5.5 from days 0 to 6 to inhibit methanogenesis and induce acetogenesis. On the 7th day, 2 mM of BES was added to inhibit methanogenesis and a pH of 6.2 was maintained to induce acetogenesis. On the 21st day of bioreactor operation, 1 g/L of YE and 0.9 g/L of L-cysteine-HCl were added for inducing growth of biomass, as it was low originally and thus allowing to reach a redox potential of -189 mV which was found to be suitable for CO metabolizing biomass to grow and simultaneously produce acids. On the 31st day, the pH was changed to 4.9 to stimulate solventogenesis.

4.2.3 Batch studies on syngas utilization by the enriched sludge

The enriched sludge and dissolved biomass collected from the reactor at the end of the experiment (46th day) were studied in the presence of syngas (mixture of CO, CO_2, H_2 and N_2 in the ratio 20:20:10:50) to check the production of metabolites at initial pH 6.2, a temperature of 33°C and 120 rpm. The pH was not controlled in the batch bottles, but the initial pH was used as it is favorable for biomass growth and production of acids. The experiments were carried out in three 100 ml bottles with 40 ml medium and 10% inoculum. After inoculation, the bottles were sparged with syngas. The overhead pressure was 1 bar. The composition of the medium was described earlier (Fernández-Naveira et al., 2016a).

4.2.4 Analytical methods

4.2.4.1 Gas-phase CO and CO$_2$ concentrations

Gas samples (1 mL) were taken from the outlet sampling ports of the bioreactor to monitor the CO, CO_2 and CH_4 concentrations. A HP 6890 gas chromatograph (GC, Agilent Technologies, Madrid, Spain) equipped with a thermal conductivity detector (TCD) was used for measuring the gas-phase concentrations. The GC was fitted with a 15-m HP-PLOT Molecular Sieve 5A column (ID - 0.53 mm; film thickness - 50 μM). The initial oven temperature was kept constant at 50 °C, for 5 min, and then raised by 20 °C/min for 2 min, to reach a final temperature of 90 °C. The temperature of the injection port and the detector were maintained constant at 150 °C. Helium was used as the carrier gas at a flow rate of 2 mL/minute. Another HP 5890 gas chromatograph (GC, Agilent Technologies, Madrid, Spain) connected with a TCD was used for measuring CO_2 and CH_4. The injection, oven, and detection temperatures were maintained at 90, 25, and 100 °C, respectively. The area obtained from the GC was correlated with the concentration of the gases.

4.2.4.2 Fermentation products

The water-soluble products, acetic acid, butyric acid, lactic acid, propionic acid, hexanoic acid, ethanol, butanol and hexanol", were analysed from liquid samples (1 mL) taken about every 24 h from the bioreactor medium using an HPLC HP1100 (Agilent Technologies, Madrid, Spain) equipped with a Supelcogel C610 column and a UV detector at a wavelength of 210 nm. The mobile phase was a 0.1 % ortho-phosphoric acid solution fed at a flow rate of 0.5 mL/min. The column temperature was set at 30 °C. Before analysing the concentration of the water-soluble products by HPLC, the samples were centrifuged (7000g, 3 min) using a bench-scale centrifuge (ELMI Skyline ltd CM 70M07, Riga Latvia).

4.2.4.3 Redox potential and pH measurement

In the bioreactor, an Ag/AgCl reference electrode connected to a transmitter (M300, Mettler Toledo, Inc., Bedford, MA, USA) was used to measure the redox potential. An on line pH controller (Hamilton, USA) was also placed in the bioreactor in order to maintain a constant pH during its operation. The pH was measured on-line with an (in built) sensor and adjusted using either 1M NaOH or HCl solutions.

4.2.4.4 Measurement of biomass and specific growth rate of the solventogenic acetogens

An UV-visible spectrophotometer (Hitachi, Model U-200, Pacisa & Giralt, Madrid, Spain) was used to measure the optical density (OD $_{600\,nm}$) of 1 mL of liquid samples withdrawn periodically from the bioreactor.

4.3 Results and Discussion

4.3.1 Sludge enrichment and production of acids and alcohols

An initial relatively low pH of 5.5 was first applied to the continuous gas-fed stirred tank bioreactor in order to convert CO and the residual acids of the inoculum into alcohols, as acidic conditions have been reported to favour the production of solvents (e.g., alcohols) in gas fermentation (Abubackar et al., 2012; Kennes et al., 2016). The pH was maintained constant through automatic pH regulation. However, during the first week of operation, there was no production of alcohols at this pH value (pH 5.5), though there was some methane (CH_4) production as shown in Figure 4.S1 reaching about 0.7 g/m3 on 7th day and subsequently dropped to zero.. Production of H_2 was not observed in the bioreactor. It was probably readily consumed by the anaerobic microbial community present in the inoculum.

Since the goal of this study was to enrich for gas fermenting anaerobic alcohol producing bacteria rather than for methanogens, 2 mM BES was added to the bioreactor medium on the 7th day, which completely inhibited methanogenesis. Simultaneously, there was a slight increase in the ethanol production up to ~ 150 mg/L (Figure 4.1a). The lag phase continued till the 21st day with hardly any detectable biomass increase, measured in terms of optical density (Figure 4.1b). During this initial period, nutrients, trace metals and vitamins were not added to the medium, as mixed populations in sludges generally require less rich media than pure cultures for their growth and activity (Kafkewitz et al., 1999). However, syngas or CO fermenting acetogens may require some additional nutrients as most studies reported that vitamins are required, mainly for the growth and metabolism of pure cultures (Fernández-Naveira et al., 2016a, 2017a, b, c, d). Because of this low activity and long lag phase, YE and L-cysteine-HCl were added on the 21st day as YE is expected to induce biomass growth and hence improve the production of acids and, later on, presumably their bioconversion to alcohols. Besides, 0.9 g/L L-cysteine-HCl, which was higher than the previous amount of 0.6 g/L of L-cysteine-HCl, was added in order to lower the redox potential of the bioreactor medium. Complete absence of YE in the culture medium of *C.carboxidivorans* has been found to produce butanol (1.0 g/L), hexanol (0.9 g/L) and ethanol (3.0 g/L) with a lag phase of less than 2 days in the presence of syngas as the sole substrate (Philips et al., 2015). But, in case of mixed cultures, one of the reasons for the longer lag phase of the solventogenic acetogens may be due to the presence of CO as the sole carbon substrate, which is more difficult to convert compared to when it is present in a mixture with H_2 (in lesser percentage) as in the case of syngas.

Wan et al. (2017) reported that in the absence of YE, there was limited growth of the syngas fermenting strain *C. carboxidivorans* P7 with a lag phase of 5 days, even in the presence of easily consumable sugars like glucose and fructose. The presence of YE has also been found to improve the availability of essential amino acids in *C.carboxidivorans* P7, while these amino acids, particularly aspartate and glutamate, were not synthesized *de novo*. Low YE concentrations or its absence have been observed to induce solventogenesis (Abubackar et al., 2012), while the initial presence of YE would be required in order to stimulate cell synthesis and growth of gas fermenting clostridia. Its addition in this study was thus mainly aimed at stimulating the required initial biomass growth, as shown in Figure 1b, rather than solventogenesis. There are studies on the production of alcohols and acids from CO by pure cultures by replacing YE with corn steep liquor (Maddipati et al., 2011, Liu et al., 2014a, b), e.g. a pure culture study with *Clostridium* strain P11 produced 6.1 g/L of ethanol from syngas in the presence of YE, and still 1.7 g/L more when corn steep liquor was present in the medium (Maddipati et al., 2011). In comparison to all reported studies, this study could produce hexanol and butanol using CO as the sole substrate in the reactor study (Figure 4.1a). Batch experiments were also performed with syngas in order to find out the capability of the enriched culture to use H_2 and CO_2 as the substrate (Figure 4.3). In the presence of CO, this is the sole carbon substrate used for growth and as electron donor, but in case of syngas, H_2 is also used as electron donor (Bertsch et al., 2015). Moreover, no vitamins were supplied to the medium which can be a reason for the longer lag phase. YE was primarily used in this study for enrichment as it is mostly used in CO fermentation (Fernández-Naveira et al., 2016 a). Future research is required on decreasing the required YE concentration and replacing it with alternative nitrogen sources.

L-cysteine-HCl, a non-essential amino acid, is reported to be a crucial component to decrease the redox potential of anaerobic systems and stimulate cell growth (Sim et al., 2008). A more reduced medium or the presence of a higher concentration of reducing agents such as L-cysteine-HCl have been reported to improve solventogenesis and alcohol production in *C. autoethanogenum* (Abubackar et al., 2012). The redox potential kept gradually decreasing down to a minimum value of -400 mV on the 28th day of operation and a typical sigmoid bacterial growth was then observed during that period, as inferred from the OD_{600} values (Figure 1b).

The production of acetic acid and ethanol was observed (Figure 1a) from the 23rd day onwards. The highest amounts of acetic acid and ethanol produced were 6.18 g/L and 11.1 g/L on the 29th and 44th day, respectively, confirming that the acid was produced first to be later converted into the alcohol. The ethanol, butanol and hexanol concentrations detected in the fermentation broth on day 25 were 439, 91 and 400 mg/L, respectively, indicating the initiation of solventogenesis in the fermentation medium already at pH values of 6.2. Indeed, although only alcohols are usually produced at low pH in solventogenic anaerobic bacteria, both acids as well as alcohols can be found at higher pH values (6.2) by many gas fermenting acetogens (Fernández-Naveira et al., 2017a).

From Figs. 4.1a and 4.1b it is evident that once biomass growth stopped there was a net fast decrease of the volatile fatty acids concentrations with a similarly fast increase in the concentrations of the corresponding alcohols (C2, C4, and C6). Nevertheless, before biomass growth completely ceased, the volatile fatty acid accumulation rates decreased and levelled off with the production of alcohols during that period. This indicates that at the end of the high pH period, solventogenesis can already start. Butanol and hexanol were produced later than ethanol and somewhat more slowly. Ethanol appears generally first in anaerobic gas fermentation processes, while the production of higher alcohols occurs at a later stage (Philps et al., 2015, Fernández-Naveira et al., 2017b, c, Ukopong et al., 2012).

The highest amounts of butanol and hexanol were 1.8 and 1.46 g/L, detected on the 41st and 42nd day, respectively. Finally, the highest production of butyric acid and hexanoic acid was 1.33 and 0.42 g/L on the 25th day and 42nd day, respectively (Figure 1a). However, the butyric and hexanoic acids were consumed later on and rapidly disappeared. The lower production of those longer chain organic acids was partly due to their conversion into alcohols. Overall, it can thus be concluded that the applied strategy, based mainly on evaluating the effect of medium composition and alternating its pH, was successful in the enrichment of solventogenic anaerobic populations.

Table 4.1 summarizes the approximate CO utilisation for the production of different metabolites in the enrichment phase. It gives an overview of the efficiency of conversion including the startup period comprising of methanogenesis and the lag phase. Due to dissolved organic substrates in the sludge inoculum, actual CO conversion to the products could not be predicted. Including the startup phase 13.9% conversion of CO was achieved. The overall conversion of CO has been calculated at the end of the enrichment phase with respect to the yield of alcohol as it was the final product.

Phases	CO passed through the medium (g)	Mole of CO passed through the medium	Equations for soluble metabolites	Highest concentration of metabolites	Theoretical CO utilised	Moles of CO utilised	
Methanogenic phase (0-7)	109.71	3.91	$CO+3H_2 \rightarrow CH_4 +H_2O$ (probable) $C_3H_6O_3 \rightarrow 3CH_4$ $HCOOH \rightarrow CH_4$	Could not be estimated as the inoculum contained residual substrates (acids mainly) that can also yield CH_4	-		
Lag phase (8-20)	203.69	7.27					
Enrichment phase (21-46)	412.42	14.72	$6CO+3H_2O \rightarrow C_2H_5OH + 4CO_2$	11.1 g/L ethanol	40.5 g	1.44	
			$12CO+5H_2O \rightarrow C_4H_9OH+8CO_2$	1.8 g/L butanol	9.8 g	0.35	

		$18CO+7H_2O \rightarrow$ $C_6H_{13}OH+12CO_2$	1.46 g/L hexanol	7.17 g	0012
					2.04
		Biomass potentially produced from CO was unaccounted for		Cumulative 57.47 g (13.5%)	

Table 4.1 CO utilisation for the production of different metabolites in the enrichment phase

Figure 4.1c shows that some lactic acid, formic acid and propionic acid were detected at low concentrations ($\leq$ 200 mg/l), but both their production and subsequent consumption took place during the high pH stage. Lactic acid was mainly present during the first two weeks, reaching a maximum concentration of 187.2 mg/L on the 11th day. Formic acid had an initial concentration of 98 mg/L, provided by the inoculum. It was completely utilised during the initial phase, suggesting that formic acid was used as the substrate for methanogenesis along with CO. However, on suppression of methanogenesis, during the 2nd week of operation, formic acid was found to sharply increase again to values of ~ 120 mg/L on the 7th-8th day, but was then readily consumed again during the 2nd week of operation. With the onset of acetogenesis on the 21st day, the formic acid concentration was again found to transiently increase up to 200 mg/L on the 24th day. Thereafter, from day 25 onwards, the formic acid concentration decreased again and it was completely consumed on the 31st day of operation. The profile of propionic acid fluctuated during the course of bioreactor operation with a somewhat similar pattern as that of formic acid. The concentration of propionic acid increased from 0 to 63 mg/L during the 1st week of operation. Later on, the suppression of methanogenesis increased the concentration of propionic acid up to 123 mg/L on the 11th day, but it again started decreasing to 20.4 mg/L on the 18th day of operation. On the 2nd day and 28th day, the measured propionic acid concentrations were 110 mg/L and 178 mg/L, which were relatively higher than its detected concentrations during the course of experiment. But the propionic acid concentration decreased on day 29, no more propionic acid was detected in the enriched medium anymore from day 30 onwards (acetogenesis has started). All those metabolites were, in any case, produced at low concentrations compared to even carbon-numbered fatty acids and had completely disappeared before solventogenesis started or when lowering the pH value. During CO fermentation, the production of these acids is significant as they can be co-metabolites, even produced in minor amounts). Future experiments can be performed using varying concentrations of these acids, while determining their effect on the production of other metabolites.

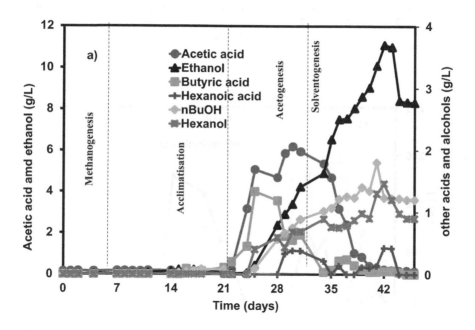

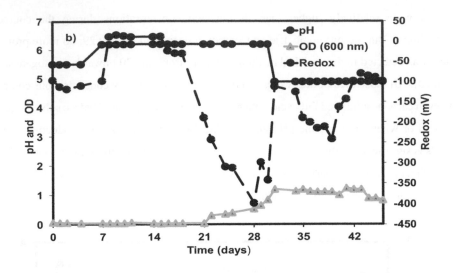

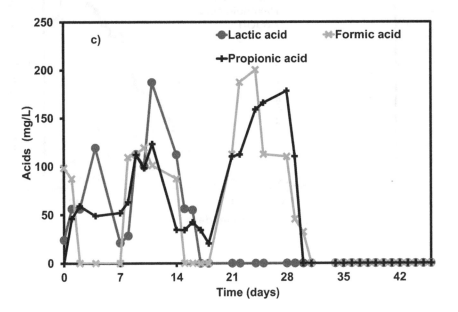

Fig. 4.1: Time evolution of parameters during enrichment: a) production of major acids and alcohols, b) profile of pH, redox potential and biomass and c) profile of other organic acids

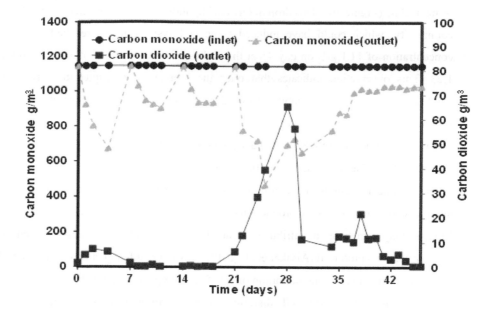

Figure 4.2: Profile of CO and CO₂

4.3.2 Biomass profile of the enriched culture of solventogenic acetogens and specific growth rate

Figure 1b shows the steady increase in the biomass concentration in the liquid phase from the 21st day of operation, after an initial lag phase. The first reading of OD was 0.065 and it then dropped down to 0.04 on the 2nd day. The subsequent optical density (OD) readings decreased further and dropped down to 0.03 on the 21st day of operation (Figure 1b). This suggested that probably the medium composition was rather unfavourable to biomass growth, as parameters like pH can be manipulated to stimulate CO utilization by growth of solventogenic acetogens (Fernández-Naveira et al., 2017d). During the following acidogenic phase, the measured OD increased to 1.2 on the 31st day. Then, on application of a low pH to induce solventogenesis, the OD started dropping somewhat, down to 1.14 on the 32nd day, to finally stabilize around 1.0 on the 40th day.

Acidogenic conditions are known to be related to biomass growth and ATP (adenosine triphosphate) generation, while it is not the case with solventogensis (Abubackar et al., 2011). Acidification resulted almost instantaneously in the termination of biomass growth or even some decline (Figure 1b). This drop in OD reading indicates a sudden shock to the bacterial community, which could have induced some cell lysis. Ethanol reached its highest concentration of 11.1 g/L on day 41 at the highest biomass concentration with an OD of 1.23. This observation indicates that once the biomass got adapted to solventogenic conditions, the metabolic capability to produce solvent-metabolites (ethanol in this case) was activated. The logarithmic phase of solventogenic bacteria with a clear exponential phase is shown in Figure 4.S2.a) and a separate graph showing only the exponential phase with Figure 4.S2.a.1) having a R^2 of 0.93 is shown.

After reaching a threshold concentration of 11.1 g/L ethanol, the cells started decaying with simultaneous reduction in the production of solvent-metabolites. The decay of the bacterial biomass may sometimes be attributed to the effect of solventogenic stress induced by the organic solvents produced (Alsaker et al., 2010). However, although a concentration of 11.1 g/L ethanol may exert some inhibition on gas fermenting clostridia, it has been shown that inhibition and eventually cell disruption can be limited at such modest ethanol concentrations (Fernández-Naveira et al., 2016b). Though the highest concentrations of butanol and hexanol, i.e., 1.18 g/L (41st day) and 1.46 g/L (42nd day), were achieved during the solventogenic stage, the concentrations of these alcohols dropped somewhat later for an unknown reason, which could have involved some butanol and hexanol degradation of , although this was not confirmed. Being more hydrophobic and hence more toxic to the bacterial cells, butanol and hexanol may perhaps also level off to a concentration that is non-toxic to the biomass (Willbanks et al., 2017). Microbial community analysis may give a better idea about the metabolism and operating strategies for optimum production of alcohols during CO fermentation.

4.3.3 Profile of CO consumption and CO_2 production

The profile of CO in both the influent and effluent gas and the CO_2 production and consumption profiles are shown in Figure 4.2. These profiles are related to the production of acids and alcohols. The initial influent and effluent CO concentration (start-up, day 0) was 1145 g/m^3. The effluent gas concentration was then found to already drop down to 923 g/m^3 of CO on the 1st day of operation to subsequently further decrease to 672 g/m^3 on the 4th day.

The addition of BES immediately decreased CO consumption and, for the next 2 weeks, there was only 10-15% CO metabolized. The effluent CO concentration fluctuated between

1145 and 1017 g/m^3 from the 7th to 21st day of operation. The minimum concentration of effluent CO was 465.8 g/m^3 (removal efficiency of 59.3 %) on the 28th day of operation to again increase to 695 g/m^3 on the 29th day of operation. The minimum redox value of -400 mV was observed on this day (Figure 4.1b) and a sharp increase in the ethanol production started from this day of operation (Figure 4.11a). This indicates that a maximum CO removal efficiency of 59 % is achieved, after which solventogenesis is accelerated. The effluent CO concentration increased again to around 1100 g/m^3 on the 36th day and remained approximately constant on the 46th day of operation.

The production of CO_2 (Figure 4.2) was observed from the start of the operation already, indicating the active metabolism of the anaerobic sludge. During the 1st week of operation, CO_2 in the effluent increased from 1.56 to 6.23 g/m^3 to later decrease to 1.76 g/m^3 on the 7th day of operation. From the end of the 1st week of operation till the 20th day, the effluent CO_2 concentration remained below 1 g/m^3. This was in agreement with the low CO consumption, low biomass growth and reduced activity in general during that initial phase. On the 21st day, the concentration of effluent CO_2 increased again up to 6.23 g/m^3 and reached its maximum value of 65 g/m^3 on the 27th day of operation, corresponding to the maximum CO consumption. This can be easily understood, as CO_2 is a co-product in some conversion reactions of CO into acids or alcohols, while CO_2 would usually be used up later on (Fernández-Naveira et al., 2017d). On days 28-29, the effluent CO_2 concentration dropped to 56 and 11.34 g/m^3 consecutively (Figure 4.2). Simultaneously, biomass reached its maximum concentration as well as some metabolites, such as acetic acid and the effluent CO concentration was low (i.e., high consumption). The effluent concentration of CO_2 was around or below 11-12 g/m^3 between days 30 and 37 of operation, probably due to low CO consumption. Between days 39 and 46, the CO_2 concentration in the effluent decreased rapidly. A low value of only 2.67 g/m^3 was detected on day 46 (Figure 4.2).

4.3.4 Operating pH and redox potential

A low pH is expected to favour solventogenesis (Abubackar et al., 2012). However, using a low pH during bioreactor start-up may impede good biomass growth. Nevertheless, the operating pH during the 1st week of operation was maintained at a relatively low value of 5.5 in order to try to suppress the production of CH_4 and to convert the residual acids in the sludge/inoculum into alcohols. Such low pH was, however, not effective enough to suppress methanogenesis, requiring the addition of BES. Thereafter, the pH was increased to 6.2 and maintained at this value till the 31st day of operation, when the pH was again changed to 4.9

and then kept constant till the 46th day to stimulate solventogenesis and favour the conversion of acids, produced at high pH of 6.2, into alcohols (Figure 4.1a).

The redox potential of the fermentation broth changed from -96 mV to -110 mV during the 1st week of operation. From the 8th day of operation, after adding BES, the redox value became positive with a value of 11-12 mV, indicating inhibition of the metabolism. From the 15th day, the redox value again became negative and reached values of -23.9 and -30.0 mV between days 16 and 18. From the 21st day, the redox value dropped to -189 mV and kept decreasing steadily to -342 mV when reaching the 31st day, i.e. during the acetogenic phase. The drop of the redox potential (Figure 4.1b) was accompanied by the sudden growth of biomass (Figure 4.1b), concomitant to a significant gas consumption (Figure 4.2) and metabolite production. Changes in redox potential are known to be affected by the pH value and the metabolic activity. From day 40, the redox value increased to -162.3 mV, and thereafter it continued increasing steadily from the 41st day up to the 45th day to reach a value of -89.23 mV. The redox value at the end of the bioreactor operation, on the 46th day, was -98.14 mV (Figure 4.1b).

4.3.5 Syngas as potential substrate mixture for the enriched anaerobic sludge

CO metabolizing acetogens may also be able to grow on H_2/CO_2 or syngas mixtures. The ability to metabolize such mixtures is interesting and relevant from a practical point of view (Rachbauer et al., 2017). Bottle experiments showed that the CO enriched sludge is indeed also able to use the mixture of CO /CO_2/H_2 as carbon and energy source. The highest concentrations of the metabolites were recorded and reached the following values: acetic acid, 1.9 g/L, ethanol, 0.89 g/L, butyric acid, 0.98 g/L, butanol, 0.6 g/L and hexanoic acid, 0.34 g/L (Figure 4.3a). H_2 and CO were completely utilised after 7 days batch experiment, while there was still CO_2 as a major metabolic co-product of syngas fermentation and as measured from the headspaces of the bottles (Figure 4.3b). Since these are bottle experiments, pH was not regulated and was allowed to drop freely as a result of acids production, reaching a final pH value of 5.1. The logarithmic exponential growth of the biomass is shown in Figure 4.S2.a.1) with an exponential phase from 2 to 8 days of operation with a R^2 value of 0.96.

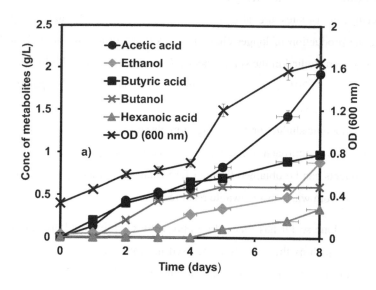

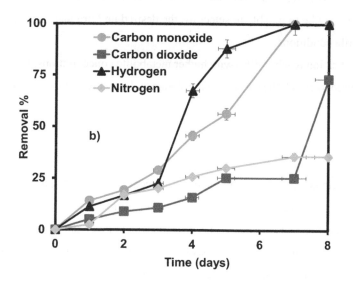

Fig. 4.3: Studies of syngas fermentation by enriched sludge at initial pH of 6.2: a) profile of production of metabolites and growth of biomass, b) profile of consumption of syngas components.

4.3.6 Practical application and future perspectives

Only very few acetogenic isolates are able to produce (bio)ethanol through C1 gas fermentation, while the production of longer chain alcohols (butanol and hexanol) is still scarcer and has mainly been studied in the single species *C. carboxidivorans* (Philips et al., 2015, Fernández-Naveira et al., 2017a, b, c). Enrichment of additional microbial populations is therefore highly interesting. This study shows the possibility to enrich for solventogenic acetogens from a methanogenic sludge for producing biofuels and alcohols such as ethanol, butanol, and hexanol, as an approach for a potential larger scale production of such compounds. These products can be obtained through fermentation of waste gas emissions or from syngas produced from waste or renewable resources.

Although most research done so far has focused on pure cultures, mixed cultures present some interesting features such as their ability to be used in bioreactors under non sterile conditions, their generally lower requirements for micronutrients, their good resistance to inlet shock loads or starvation compared to single strains or their higher resistance to toxic or inhibitory conditions. Working under non sterile conditions reduces operating costs. Conversely, one issue would be to be able to select for the desired end products whenever working under non sterile conditions.

In this study, the CO utilisation reached 59% with higher metabolite concentrations, though a longer time was required to enrich the solventogenic sludge and activate CO metabolizing microorganisms,

4.4 Conclusions

Sequential selective enrichment of solventogenic acetogens from a highly active methanogenic sludge was possible, with methane production during the 1st week and a prolonged lag phase for acetogens (2nd week). Acetogenesis was induced upon addition of YE (yeast extract) and L-cysteine-HCl, with subsequent solventogenesis. The successful enrichment strategy first allowed the growth of acetogens, followed by acidification which ended biomass growth and stimulated solventogenesis. 11.1 g/L ethanol was obtained, which is higher than most values reported in pure culture studies under similar conditions. Mixed cultures may produce higher concentrations of alcohols, though a longer lag phase may be required for stimulating solventogens. This study shows that it is possible to stop the bioconversion steps at the level of solvent (alcohols) production by acetogens, while limiting or inhibiting the activity of methanogens, which are more sensitive to non-optimal operating conditions besides exhibiting low growth rates.

References

Abubackar, H. N., Veiga, M.C., Kennes, C. Biological conversion of carbon monoxide: rich syngas or waste gases to bioethanol. Biofuels, Bioprod and Bioref. 5 (2011): 93-114.

Abubackar, H. N., Veiga, M.C., Kennes, C. Biological conversion of carbon monoxide to ethanol: effect of pH, gas pressure, reducing agent and yeast extract. Bioresour. Technol. 114 (2012): 518-522.

Abubackar, H. N., Veiga, M.C., Kennes. C. Carbon monoxide fermentation to ethanol by *Clostridium autoethanogenum* in a bioreactor with no accumulation of acetic acid. Bioresour. Technol. 186 (2015): 122-127.

Abubackar, H. N., Fernández-Naveira, A., Veiga, M.C., Kennes, C. Impact of cyclic pH shifts on carbon monoxide fermentation to ethanol by *Clostridium autoethanogenum*. Fuel. 78 (2016): 56-62.

Alsaker, K.V., Paredes, C., Papoutsakis, E.T. Metabolite stress and tolerance in the production of biofuels and chemicals: gene-expression-based systems analysis of butanol, butyrate, and acetate stresses in the anaerobe *Clostridium acetobutylicum*. Biotechnol. Bioeng. 105 (2010): 1131-1147.

Bengelsdorf, F. R., Straub, M., Dürre, P. Bacterial synthesis gas (syngas) fermentation. Environ. Technol. 34 (2013) 1639-1651.

Bertsch, J., Müller V. Bioenergetic constraints for conversion of syngas to biofuels in acetogenic bacteria. Biotechnol biofuels. 8 (2015): 210

Devarapalli, M., Atiyeh, H.K., Tanner, R.S., Philips, J. R., Saxena, J., Lewis, R.S., , M.R., Huhnke, R.L. Ethanol production during semi continuous syngas fermentation in trickle bed reactor using *Clostridium ragsdalei,* Bioresour. Technol. 209 (2016): 56-65.

Fernández-Naveira, Á., Abubackar, H.N., Veiga, M.C., Kennes, C. Efficient butanol-ethanol (BE) production from carbon monoxide fermentation by *Clostridium carboxidivorans.* Appl. Microbiol. Biotechnol. 100 (2016a): 3361-3370.

Fernández-Naveira, Á., Abubackar, H.N., Veiga, M.C., Kennes, C. Carbon monoxide bioconversion to butanol-ethanol by *Clostridium carboxidivorans*: kinetics and toxicity of alcohols. Appl. Microbiol. Biotechnol. 100 (2016b): 4231-4240.

Fernández-Naveira, Á., Veiga, M.C., Kennes, C. H-B-E (hexanol-butanol-ethanol) fermentation for the production of higher alcohols from syngas/waste gas. J. Chem. Technol. Biotechnol. 92 (2017a): 712-731.

Fernández-Naveira, Á., Veiga, M.C., Kennes, C. Production of chemicals from C1 gases (CO, CO_2) by *Clostridium carboxidivorans*. World J. Microbiol. Biotechnol. 33 (2017b): 43-53.
Fernández-Naveira, Á., Veiga, M.C., Kennes, C. Glucose bioconversion profile in the syngas-metabolizing species *Clostridium carboxidivorans*. Bioresour. Technol. 244 (2017c): 552-559.

Fernández-Naveira, Á., Veiga, M.C., Kennes, C. Effect of pH control on the anaerobic H-B-E fermentation of syngas in bioreactors. J. Chem. Technol. Biotechnol. 92 (2017d): 1178-1185.

Fillat Ú., Ibarra, D., Eugenio, Moreno, M. E., Tomás-Pejó, E., Martín-Sampedro, R. Laccases as a potential tool for the efficient conversion of lignocellulosic biomass: a review. Fermentation, 3(2017): 17-46.

Grant, C.L., Pramer, D. Minor element composition of yeast extract. J. Bacteriol. 84 (1962) 869-870.

Huhnke, R.L., Lewis, R.S., Tanner, R.S. Oklahoma State University and University of Oklahoma, 2010. Isolation and characterization of novel clostridial species. U.S. Patent 7,704,723.

Kennes, D., Abubackar, H.N., Diaz, M., Veiga, M.C., Kennes, C. Bioethanol production from biomass: Carbohydrate vs syngas fermentation. J. Chem. Technol. Biotechnol. 91 (2016): 304-317.

Kafkewitz, D., Togna, M.T. The methods of anaerobic microbiology with emphasis on environmental applications. Int. J. Env. Studies 56 (1999): 141-170.

Köpke M., Mihalcea, C., Liew, F., Tizard, J.H., Ali, M.S., Conolly, J.J., Al-Sinawi, B., Simpson., S.D. 2,3-Butanediol production by acetogenic bacteria, an alternative route to chemical synthesis, using industrial waste gas. Appl. Environ. Microbiol. 77 (2010): 5467-5475.

Lagoa-Costa, B., Abubackar H.N., Fernández-Romasanta, M., Kennes, C., Veiga, M.C. Integrated bioconversion of syngas into bioethanol and biopolymers. Bioresour. Technol. 239 (2017): 244-249.

Liu, K., Atiyeh, H.K., Stevenson, B.S., Tanner, R.S., Wilkins, M.R., Huhnke, R.L. Mixed culture syngas fermentation and conversion of carboxylic acids into alcohols. Bioresour. Technol. 152 (2014a): 337-346.

Liu, K., Atiyeh, H.K., Stevenson, B.S., Tanner, R.S., Wilkins, M.R., Huhnke, R.L. Continuous syngas fermentation for the production of ethanol, n-propanol and n-butanol. Bioresour. Technology, 151 (2014b): 69-77.

López, M.E., Rene, E.R., Veiga, M.C., Kennes, C. Chapter 9: Biogas Technology and Cleaning Techniques, in: Lichtfouse, E., Schwarzbauer, J., D. Robert., D. (Eds.) (2012), Environmental Chemistry for a Sustainable World, Springer, Germany, pp. 347-377.

López, M.E., Rene, E.R., Veiga, M.C., Kennes, C. Chapter 13: Biogas Upgrading. In: C. Kennes and M.C. Veiga (eds.) (2013). Air Pollution Prevention and Control: Bioreactors and Bioenergy, J. Wiley & Sons, Chichester, United Kingdom, pp. 293-318.

Maddipati, P., Atiyeh, H.K., Bellmer, D.D., Huhnke, R.L. Ethanol production from syngas by Clostridium strain P11 using corn steep liquor as a nutrient replacement to yeast extract. Bioresour. Technol. 102 (2011): 6494-6501.

Philips, J.R., Atiyeh, H.K., Tanner, R.S., Torres, J.R., Saxena, J., Wilkins, M.R., Huhnke, R.L. Butanol and hexanol production in *Clostridium carboxidivorans* syngas fermentation: medium development and culture techniques. Bioresour. Technol. 190 (2015): 114-121.

Rachbauer, L., Beyer, R., Bochmann, G., Fuchs, W. Characteristics of adapted hydrogenotrophic community during biomethanation. Sci. Tot. Environ. 595 (2017): 912-919.

Ramió-Pujol S., Ganigué R., Bañeras, L., Colprim, J. Incubation at 25° C prevents acid crash and enhances alcohol production in *Clostridium carboxidivorans* P7. Bioresour. Technol. 192(2015): 296-303.

Sancho Navarro, S., Cimpoia, R., Bruant, G., Guiot, S.R. Biomethanation of syngas using anaerobic sludge: shift in the catabolic routes with the CO partial pressure increase. Front. Microbiol. 7 (2016): 1188-1200.

Sim. J.H., Kamaruddin, A.H. Optimization of acetic acid production from synthesis gas by chemolithotrophic bacterium - *Clostridium aceticum* using statistical approach. Bioresour. Technol. 99 (2008): 2724-2735.

Singla, A., Verma, B., Lal, B., Sharma, P.M. Enrichment and optimization of anaerobic bacterial mixed culture for conversion of syngas to ethanol. Bioresour. Technol. 172 (2014): 41-49.

Shen, N., Dai, K., Xia, X.Y., Zeng, R.J., Zhang, F. Conversion of syngas (CO and H_2) to biochemicals by mixed culture fermentation in mesophilic and thermophilic hollow-fiber membrane biofilm reactors. J. Cleaner Prod. 202 (2018): 536-542.

Omar, B., Abou-Shanab, R., El-Gammal, M., Fotidis, I.A., Kongins, P.G., Zhang, Y., Angelidaki, I. Simultaneous biogas upgrading and biochemical production using anaerobic bacterial mixed culture. Water Res. 142 (2018): 86-95.

Ukopong, M.N., Atiyeh, H.K., De Lorme, M.J., Liu, K., Zhu, X., Tanner, R.S., Wilkins, M.R., Stevenson, B.S. Physiological response of *Clostridium carbioxidivorans* during conversion of synthesis gas to solvents in a gas-fed bioreactor. Biotech Bioeng. 109 (2012): 2720-2728.

Van Groenestijn, J.W., Abubackar, H.N., Veiga, M.C., Kennes, C. Chapter 18: Bioethanol. In: C. Kennes and M.C. Veiga (eds.) (2013). Air Pollution Prevention and Control: Bioreactors and Bioenergy, J. Wiley & Sons, Chichester, United Kingdom. pp 431-463.

Van Lier, J.B., Van Der Zee, F.P., Frijters, C.T.M.J., Ersahin, M.E. Celebrating 40 years anaerobic sludge bed reactors for industrial wastewater treatment. Rev. Environ. Sci. Biotechnol. 14 (2015): 681-702.

Wan, N., Sathish, A., You, L., Tang, Y.J., Wen, Z. 2017. Deciphering *Clostridium* metabolism and its responses to bioreactor mass transfer during syngas mass transfer during syngas fermentation. Sci. Rep. 7, 10090-10100.

Wang, J., Wan, W. Kinetic models for fermentative hydrogen production: a review. Int. J. Hydrogen Energy 34 (2009): 3313-3323.

Webster, T.M., Smith, A.L., Smith, R. R., Pinto, A.J., Pinto, K.F., Raskin, L. Anaerobic microbial community response to methanogenic inhibitors 2-bromoethanesulfonate and propionic acid. Microb. Open. (2016): 537-550.

Wilbanks, B., Trinh, C.T. Comprehensive characterization of toxicity of fermentative metabolites on microbial growth. Biotechnol. Biofuels 10 (2017): 262.

Xu, S., Fu, B., Zhang, L., Liu, H. Bioconversion of H_2/CO_2 by acetogen enriched cultures for acetate and ethanol production: the impact of pH. World J. Microbiol. Biotechnol. 31 (2015): 941-950.

Zhang, F., Ding, J., Zhang, Y., Chen, M., Ding, Z.W., van Loosdrecht, M.C., Zeng, R.J. Fatty acids production from hydrogen and carbon dioxide by mixed culture in the membrane biofilm reactor. Water Res. 47 (2013): 6122–6129.

Supplementary materials

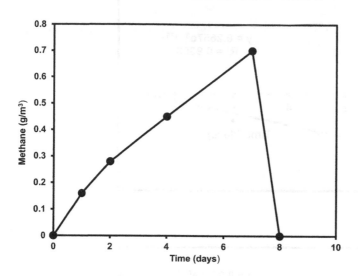

Figure 4.S1.Production of methane in the 1ˢᵗ week of operation

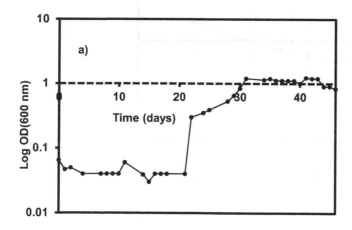

Figure 4.S2: Logarithmic plot of OD (600 nm) a) in the reactor using CO

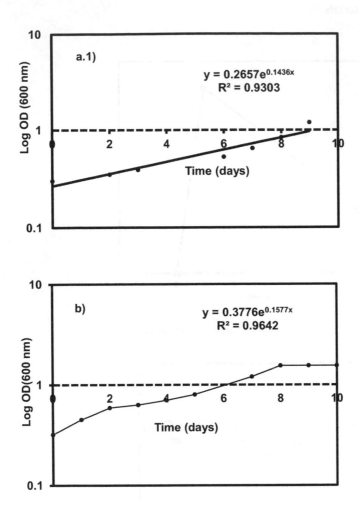

Figure 4.S2: a.1) exponential phase plot of OD from 21-31 days in the reactor using CO, b) Logarithmic plot of OD (600 nm) from enriched batch culture using syngas

Table 4.S1: Characteristics of acclimatized sludge and CO unacclimatized sludge

Characteristics	Unacclimatized sludge	CO acclimatized sludge (used in the reactor)
pH	8.6	7.9
TS (g/kg)	115.6±0.2	112.7±0.2
VS (g/kg)	99.1±0.3	94.3±0.3
Acetic acid (g/L)	0.123±0.001	0.213±0.001
Propionic acid (g/L)	0.253±0.001	0.062±0.001
Iso-butyric acid (g/L)	0.312±0.001	0.1±0.001
Butyric acid (g/L)	0.129±0.001	0.1±0.001
Iso-valeric acid (g/L)	0.012±0.001	0
Valeric acid (g/L)	0.017±0.001	0
Lactic acid (g/L)	0.5±0.001	0
Ethanol (g/L)	0.063±0.001	0.187±0.001
CH_4 (g/m^3)	0.13±0.001	0.79±0.002

Chapter 5

Effect of consecutive deficiency of selenium and tungsten on production of acids and alcohols from CO

This chapter is under preparation for submission as entitled Chakraborty, S., Rene, E.R., Lens, P.N., Veiga, M.C., Kennes, C. Effect of tungsten and selenium for CO and syngas bioconversion by enriched anaerobic sludge. (To be submitted to chemosphere)

Abstract

The effect of the lack of trace metals, namely tungsten and selenium, on the production of acids and alcohols by CO-enriched anaerobic sludge was investigated. The CO-enriched sludge was first supplied a tungsten-deficient medium (containing selenium) and, in a next assay, a selenium-deficient medium (containing tungsten) was fed to a continuous gas-fed bioreactor operating for 36 days at a CO flow rate of 10ml/min. An initial pH of 6.2 followed by a pH decrease to 4.9 yielded 7.34 g/L of acetic acid as major acid at the end of the pH 6.2 stage. Subsequently, bioconversion of the acids at a lower pH of 4.9 yielded 1.85 g/L ethanol and 1.2 g/L butanol in the absence of tungsten (tungstate). A similar follow up assay in the same bioreactor with two consecutive periods of pH maintained at 6.2 first and then decreased to 4.9, with selenium deficient medium yielded 6.6 g/L of acetic acid at pH 6.2 and 4 g/L of ethanol as well as 1.88 g/L butanol at the lower pH 4.9.The only known CO fixing organism able to produce alcohols, as revealed by the microbial community analysis, was *Clostridium autoethanogenum* which was found to be present in the bioreactor both in the tungsten and selenium deprived medium.

Keywords: carbon dioxide, carbon monoxide, tungsten, selenium, ethanol, butanol

5.1 Introduction

Production of chemicals from pollutants has become a main research interest in order to develop a circular economy. Particularly, the production of biofuels, e.g. bio alcohols, from gas, liquid or solid waste streams has never been of such importance as in the current scenario of climate change, characterized by a huge increase in per capita fuel consumption with simultaneous depletion of reserves of fossil fuels. CO, CO_2 and/or syngas (a mixture of varying composition of CO, CO_2 and H_2) fermentation is emerging as a potential alternative route for the production of bio alcohols as suitable future bio-fuels, simultaneously or alternately generating value added chemicals like volatile fatty acids. The syngas bioconversion process to ethanol is even considered for commercialization (Köpke et al., 2011, Yasin et al., 2019). Primarily, pure cultures of CO metabolizing clostridia, e.g. *Clostridium autoethanogenum*, have been exploited for the production of bio alcohols (Kennes et al, 2016). Enhanced alcohol yields, compared to acids, can be achieved by alteration of the pH (Abubackar et al., 2016), optimization of the trace metal composition and concentrations (Saxena and Tanner, 2011; Abubackar et al., 2015; Fernández-Naveria et al., 2019) or using different bioreactor design and operating conditions (Ungerman and Heindel, 2007, Munashinge and Khanal, 2010; Fernández-Naveria et al., 2017).

Defined mixed co-culture systems can be applied for syngas fermentation. For example, cultures dominated by *Alkalibacterium bacchi* strain CP15 and *Clostridium propionicum* (Liu et al. 2014a) were used for the production of propanol and butanol, or co-cultures of the species *Clostridium kluyveri* and *Methanogen* M166 for the production of caproic acid (Yan and Dong, 2018). Other reports on the enriched anaerobic sludge performing continuous syngas fermentation are the "mixed culture TERI SA1" (Singla et al., 2014) and thermophilic sludge termed T-syn (Alves et al., 2013). The most recent one is based on the performance of CO fermentation for 100 days in a novel upflow reactor inoculated with sludge from a pulp and waste paper treatment plant dominated by *Alcaligenes* and *Acinetobacter* producing up to C8 compounds (He et al., 2018).

The concentrations of the different metabolites obtained in the above mentioned studies were, however, not high and no further optimization studies have been conducted with them. Moderate concentrations of C2-C6 like 6.18 g/L acetic acid (30th day), 1.18 g/L butyric acid (28th day), and 0.423 g/L hexanoic acid (32nd day), (acids and/or alcohols) metabolites were also achieved in a continuous CO fed bioreactor by an acetogenic/solventogenic mixed culture enriched from a methanogenic sludge (Chakraborty et al., 2019). The above mentioned study was performed in the presence of all trace elements, vital for CO/syngas fermentation, according to the composition used by Fernandez Naveira et al. (2017) which would have presumably produced the highest concentrations of the metabolites. The present study is a step forward to find the production profile of metabolites from CO under two different conditions, which have not been reported before in case of mixed culture fermentation: i) when the nutrient medium is deficient of tungsten (in the form of tungstate), though selenium (in the form of selenate) is present and ii) when the nutrient medium is deficient of selenium, though tungsten is present.

Some previous studies have been performed with pure bacterial cultures, such as *Clostridium ragsdalei* (Saxena and Tanner, 2011), *Clostridium autoethanogenum* (Abubackar et al., 2015), and *Clostridium carboxidivorans* (Fernández-Naveira et al., 2019). However, the results showed contradictory conclusions, i.e., selenium stimulated solventogenesis in *Clostridium ragsdalei* (Saxena and Tanner, 2011), but not in other reported strains (Abubackar et al., 2015; Fernández-Naveira et al., 2019). Trace elements in the CO/syngas fermentation may have key roles in bacterial enzymes as, for example, FDH (formate dehydrogenase), AOFR (aldehyde oxidoformidoreductase), ADH (alcohol dehydrogenase) and H_2ase (hydrogenases). This study investigates the dependency of CO enriched anaerobic sludge of the presence of two trace elements, selenium and tungsten in the bioreactor medium.

5.2 Material and Methods

5.2.1 Biomass and medium composition

The original inoculum containing anaerobic sludge, predominated by methanogens (sampled from Biothane, Delft, The Netherlands), was first enriched in a continuous CO fed bioreactor. It produced 11.1 g/L ethanol, 1.8 g/L butanol and 1.46 g/L hexanol (Chakraborty et al, 2019). This inoculum was enriched in modified basal anaerobic medium (NaCl, 0.9 g; $MgCl_2 \cdot 6H_2O$, 0.4 g; KH_2PO_4, 0.75 g; K_2HPO_4 0.5 g; $FeCl_3 \cdot 6H_2O$, 0.0025 g; 0.1% resazurin, 0.75 g), containing trace elements including 2 mg/L of sodium selenate and 2 mg/L of sodium tungstate (Fernández-Naveira et al., 2017). 1g/L YE and 0.9 g/L L/cysteine hydrochloride, were added to the medium. The original sludge was methanogenic, while it became gradually solventogenic over a 46 days period. For further studies, the enrichment medium was completely removed from the bioreactor, and the biomass flocs and residual sludge was used. The characteristics of the residual sludge are given in Table 5.S1

5.2.2 Set-up and operation of the continuous gas-fed bioreactor

Bioreactor experiments were performed in a 2L BIOFLO 110 bioreactor (New Brunswick Scientific, Edison, NJ, USA) containing 1.2 L medium which is the working volume. A microsparger was introduced inside the bioreactor for the supply of pure CO (100 %) as the sole gaseous substrate and carbon and energy source. It was fed continuously at a flow rate of 10 mL/min using a mass flow controller (Aalborg GFC 17, Müllheim, Germany). The temperature of the bioreactor was maintained at 33°C by means of a water jacket. Four baffles were symmetrically arranged to avoid vortex formation of the liquid medium and to improve mixing

5.2.3. Bioreactor operation without tungsten

In the first assay (phase I), the biomass was supplied with a tungsten deficient medium. It contained selenium in the form of sodium selenate at a concentration of 2 mg/L, but no tungsten. The residual sludge and biomass enriched over a period of 46 days in a bioreactor in a previous study was used as the inoculum (10% V/V) for the operation of the bioreactor (Chakraborty et al, 2019). The operating pH values was initially 6.2 first, for inducing acetogenesis, subsequently lowered to 4.9 for inducing solventogenesis. Two samples (1 mL each) were collected after 24 hours and the mean value was recorded for analysis.

In the phase I of operation, in tungsten deficient operation, the bioreactor medium is continuously stirred. Thus the biomass should be homogenously distributed and that biomass adhered to the reactor was used as the inoculum for the next assay. Phase II of operation was initiated with complete withdrawal of the bioreactor medium, while fresh medium was added to the bioreactor. It contained tungsten in the form of sodium tungstate at a concentration of 2 mg/L as trace element, but no selenium (in the form of selenate). Otherwise, the same operating procedure was followed as in phase I. According to the production of acids and alcohols, the acetogenic phase lasted for 6 days and the solventogenic phase (pH of 4.9) lasted from days 7-21. Two samples (1 mL each) were collected after 24 hours and the mean value of the biomass and metabolites concentrations were recorded.

5.2.4 Batch studies with CO and syngas with or without W and Se

To obtain the highest concentration of dissolved biomass and the enriched sludge was collected from i) 11th day of operation during the first phase of experiment (without W), ii) 6th day of operation from the second phase of experiments (without Se). To obtain the biomass adapted to the deficiency of tungsten and selenium, the biomass was collected from 21st day of operation from the second phase of experiment (at the end of the experiment), for subsequent batch experiments in presence of syngas and sole CO (mixture of CO, CO_2, H_2 and N_2 at the ratio 20:20:10:50) to check the production of metabolites at initial pH 6.2, a temperature of 33°C and shaking at 120 rpm. A pH of 6.2 was selected as the initial pH value as it induces acetogenesis. The experiments were carried out in duplicates, 100 ml bottles with 40 ml medium and 10% inoculum. After inoculation, the bottles were sparged with syngas. The composition of the culture medium in each different experiment was similar to the bioreactor's medium i.e., i) all trace elements except W, ii) all trace elements except Se, iii) trace elements without W and Se. Another set of experiments was carried out in batch bottles with the same operating conditions, except that CO was sparged in the batch bottles and the nutrient medium contained the trace elements except W and Se.

5.3 Analytical methods

5.3.1 Gas-phase CO and CO_2 concentrations

Gas samples of 1 mL were taken from the outlet sampling ports of the bioreactors to monitor the CO, CO_2 and CH_4 concentrations. A HP 6890 gas chromatograph (GC, Agilent Technologies, Madrid, Spain) equipped with a thermal conductivity detector (TCD) was used for measuring the gas-phase concentrations. The GC was fitted with a 15-m HP-PLOT Molecular Sieve 5A column (ID, 0.53 mm; film thickness, 50 µM). The initial oven temperature was kept constant at 50 °C, for 5 min, and then raised by 20 °C/min for 2 min to reach a final temperature of 90 °C. The temperature of the injection port and the detector were maintained constant at 150 °C. Helium was used as the carrier gas. Another HP 5890 gas chromatograph (GC, Agilent Technologies, Madrid, Spain) equipped with a TCD was used for measuring CO_2 and CH_4. The injection, oven, and detection temperatures were maintained at 90, 25, and 100 °C, respectively. The area obtained from the GC was correlated with the concentration of the gases.

5.3.2 Soluble fermentation products

The water-soluble products, namely acetic acid, butyric acid, hexanoic acid, ethanol, butanol and hexanol were analysed from liquid samples (1 mL) taken about every 24 h from the bioreactor medium using an HPLC HP1100 (Agilent Technologies, Madrid, Spain) equipped with a Supelcogel C610 column and a UV detector at a wavelength of 210 nm. The mobile phase was a 0.1 % ortho-phosphoric acid solution fed at a flow rate of 0.5 mL/min. The column temperature was set at 30 °C. Before analysing the concentration of the water-soluble products by HPLC, the samples were centrifuged (7000g, 3 min) using a bench-scale centrifuge (ELMI Skyline ltd CM 70M07, Riga, Latvia).

5.3.3 Redox potential and measurement of pH

In the bioreactor, an Ag/AgCl reference electrode connected to a transmitter (M300, Mettler Toledo, Inc., Bedford, MA, USA) was used to measure the redox potential and to ensure that anaerobic conditions and negative redox values were maintained through the study. An online pH controller was inserted in the reactor to maintain a constant pH during the operation of the bioreactor by supplying either 2M NaOH or 2M HCl solutions via peristaltic pumps.

5.3.4 Measurement of dissolved biomass concentration

An UV–visible spectrophotometer (Hitachi, Model U-200, Pacisa & Giralt, Madrid, Spain) was used to measure the optical density OD (600 nm) of 1 mL of liquid sample withdrawn from the reactor. The OD obtained from the exponential growth phase was used to calculate the maximum specific growth rate.

5.3.5 Calculation for carbon balance

CO being a sparingly soluble gas, the CO supplied is taken as the sole carbon source for the fermentation. The CO supplied is recovered primarily in the form of acids during acetogenesis and alcohols produced directly from CO, as the volatile fatty acids are the main source of alcohols.

$$Total\ CO\ supplied\ = Flow\ rate \times CO \times number\ of\ days$$

$$Total\ CO\ utilised\ in\ terms\ of\ resource\ recovery\ in\ each\ phase$$
$$= \sum CO\ from\ each\ metabolites$$

$$Theoretical\ CO\ consumption\ for\ known\ metabolites$$

$$= \frac{weight\ of\ CO\ required\ to\ synthesize\ 1\ mole\ of\ known\ metabolites}{1\ mole\ of\ known\ metabolites}$$

$$\times\ Amount\ of\ metabolites\ produced$$

5.3.6 Microbial community analysis of the reactor biomass

DNA samples were extracted from the dissolved biomass of the reactor at the end of each operation period. 10 mL of the samples were taken to extract DNA from the initial untreated sludge inoculum of this study from the enriched culture (after 46 days of experiment performed by Chakraborty et al., 2019), the two biomass samples after performing the tungsten deficient experiment in phase I for initial 15 days and the selenium deficit experiment in next phase II for the next 21 days.

A Power Soil® DNA isolation kit (MO BIO Laboratories, Inc., USA) was used to extract DNA from the defrosted filters according to the manufacturer's instructions. A primer pair Bac357F-GC and Un907R was used for amplifying the partial bacterial 16S rRNA genes by using a T3000 thermocycler (Biometra, Germany) as described by Khanongnuch et al. (2019). According to the procedure followed by Khanongnuch et al (2019), DGGE was performed with the amplified sequences by an INGENY phorU2 × 2 – system (Ingeny International BV, GV Goes, the Netherlands). The cut bands from DGGE gel were sequenced by Macrogen (South Korea). The obtained sequences were analyzed using the Bioedit software (version 7.2.5, Ibis Biosciences, USA) and compared with the sequences available at the National Center for Biotechnology Information (NCBI) database (http://blast.ncbi.nlm.nih.gov).

5.4 Results and Discussion

5.4.1 Mixed culture CO fermentation in the absence of tungsten (phase I)

5.4.1.1 Effect of lack of tungsten on the production of metabolites in gas-fed bioreactor

The experiment was started at pH 6.2, using tungsten deficient medium with continuous CO feed. Despite the lack of tungsten, similarly as observed with some other pure cultures of acetogens (e.g. *C. carboxidivorans*) (Fernández-Naveira et al., 2019), the former conditions resulted in the production of acids, mainly acetic acid, as well as a biomass increase in phase I. During this acetogenic stage, a high maximum concentration of 7.34 g/L acetic acid was achieved on the 11th day of bioreactor operation (Figure 1a) with the simultaneous consumption of CO (Figure 1b). Exponential butyric acid production started slightly later compared to acetic acid, but both metabolites did otherwise follow a rather similar accumulation pattern. The

106

maximum amount of butyric acid produced was 3.75 g/L, also on the 11th day of operation similarly as for acetic acid, when the pH was decreased from 6.2 to 4.9 (Figure 1c). The highest acetic acid production rate was 1.43 g/L.d from day 7, while the highest rate for butyric acid was 0.88 g/L/day, illustrating the slower rate besides the slight delay in the production of the C4 acid compared to the C2 acid. Although the original enriched sludge used as inoculum produced some hexanoic acid in the previously reported enrichment study, using a complete medium with tungsten (Chakraborty et al., 2019), this was not the case here. No production of alcohols was observed during the acetogenic phase in this study either. The biomass increase took place during that acetogenic phase and the OD increased from 0.04 to 1.88 on the 8th day of operation and remained nearly constant up to the 10th day of operation (Figure 1c).

During the solventogenic phase, however, the OD dropped to 1.4, but remained then constant till the end of operation. The drop in OD may be attributed to a temporary inhibition of cell metabolism and partial cell loss, due to the onset of the production of hydrophobic alcohols during solventogenesis (Fischer et al., 2008). A clear logarithmic growth (Figure S1) was observed. Taking the logarithmic value of the OD recorded in the course of 7 days, exponential growth until reaching the highest OD of 1.89, the μ_{max} was found to be 0.55/d. Pure cultures of *Clostridium carboxidivorans* exhibited a higher growth rate of 0.100/h i.e., 2.4/d in the absence of tungsten and at constant operational pH of 6.2 in an automated bioreactor with constant gas feed (Fernández Naveira et al., 2019). This observation indicates the co-culture competition of the substrate uptake for biomass synthesis and growth. Acid production was simultaneous with the growth of biomass, although the biomass concentration levelled off, and then remained constant, somewhat before the maximum concentrations of acids were reached (Figure 1a)

On the 11th day, the pH was changed from 6.2 to 4.9 in order to stimulate solventogenesis. This resulted in instantaneous ending of the production of any acids while it initiated their consumption (Figure 1a). Simultaneously, ethanol production started and 1.16 g/L ethanol had already accumulated on the next day (day 12 of operation). The ethanol concentration kept increasing, though only slightly, and levelled off at 1.88 g/L on day 15. Simultaneously, the production of a maximum concentration of 1 g/L butanol was observed (Figure 1a). However, there was no production of hexanol, which is not surprising as no hexanoic acid was found during the acetogenic stage either. The concentration of acetic acid reached a minimum value of 7.34 g/L and did only slightly drop between days 12-15. The highest conversion rate of acetic acid to ethanol was 1.44 g/L/day on the 12th day, concomitant with the highest ethanol production, but it dropped immediately to 0.17 g/L. d after that. In terms of biomass concentration, decreasing the pH led to a small drop of the OD value (Figure 1c). In the absence of tungsten, *Clostridium carboxidivorans* was shown to produce 10.05 g/L acetic acid after 162 hour, i.e., after almost 7 days of continuous operation (Fernández Naveira et al., 2019), whereas the production rate of acetic acid is much lower in the present study, i.e., 7.34 g/L after 11 days of operation (Figure 1a). On the other hand, only 1 g/L of butyric acid was produced by *Clostridium carboxidivorans* (Fernández Naveira et al., 2019), while in the mixed culture in the present study, 3.75 g/L of butyric acid was produced indicating the flux to butyric acid production.

Optimization of the effect of some trace metal concentrations for the production of metabolites from CO/syngas has been studied in pure cultures of *Clostridium carboxidivorans* (Fernández-Naveira et al., 2019), *Clostridium ragsdalei* (Saxena and Tanner, 2011), and *Clostridium autoethanogenum* (Abubackar et al., 2015). The presence of tungsten has been shown to be crucial for the production of alcohols from acids in a 2L continuously fed stirred tank bioreactor with *Clostridium carboxidivorans* (Fernández-Naveira et al., 2019). Besides, increasing concentrations of tungsten effectively stimulated the full conversion of all accumulated acids to alcohols in *Clostridium autoethanogenum*, although selenium seemed to have no positive effect on the production of alcohols under such conditions (Abubackar et al., 2015).

In *Clostridium ragsdalei*, with CO as a substrate, when increasing the tungsten concentration from $0.681\mu M$ (0.16 mg/L) to $6.81\ \mu M$ (1.6 mg/L), the ethanol production was enhanced from 35.73 mM (1.64 g/L) to 72.3 mM (3.33 g/L) at uncontrolled pH (Saxena and Tanner, 2011). In the present study, in the absence of tungsten, only 18.9% of the acetic acid produced was converted to ethanol, as the acetic acid concentration decreased from 7.34 g/L to 5.6 g/L. Theoretically, 1.75 g/L ethanol could be obtained from such an amount of acetic acid, which is close to the experimental yield of 1.85 g/L ethanol. During the original enrichment experiment of the anaerobic methanogenic sludge (Chakraborty et al., 2019), on the onset of acetogenesis, 6.2 g/L of acetic acid was produced on the 30th day of operation (10th day of acetogenesis), compared to 5.9 g/L on 10th day of acetogenesis in this study. Removal of tungsten from the medium thus appeared to result in decreased acetogenesis.

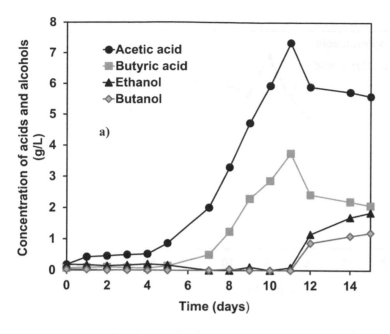

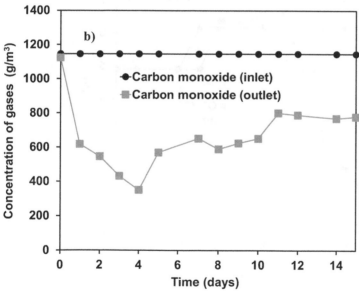

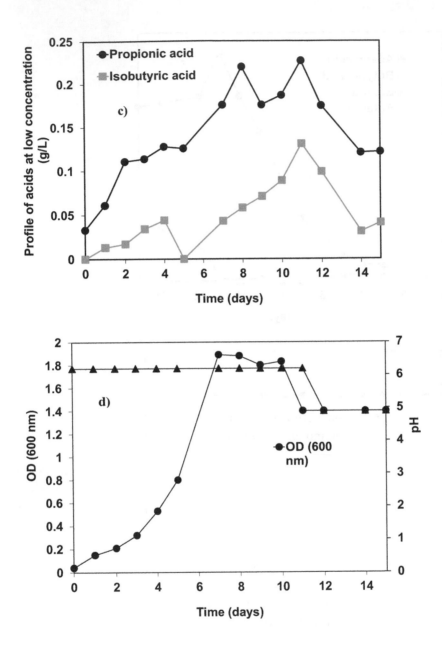

Figure 5.1 Continuous CO fed reactor experiments in the absence of tungsten (a) concentration of metabolites, (b) gas profile in continuous CO fed reactor (c) concentration of minor acids observed in the absence of tungsten and (d) OD and pH.

5.4.1.2 Detection of C3 and uncommon C4 acids in the absence of tungsten

Propionic and iso-butyric acids were two other acids present at low concentrations in the enriched sludge (Figure 1c). Propionic acid concentrations increased sharply from 33 mg/L to 220 mg/L during the acetogenic phase and decreased to 122 mg/L at the end of the solventogenic phase. The iso-butyric acid concentration increased from 0 to 141 mg/L during the acetogenic stage and decreased to 41 mg/L at the end of the solventogenesis stage. This production profile of acids was observed only in the absence of tungsten. Usually, the range of acids released to the culture medium by acetogens, at least in pure culture, is quite limited when grown on C1 gases, while a wider range has been observed in the case of carbohydrate (e.g., glucose and xylose) fermentation or mixotrophic fermentation by the same strains, e.g. *C. autoethanogenum* and *C. carboxidivorans* (Abubackar et al., 2016; Fernández Naveira et al., 2019). However, low amounts of iso-butyric acid were also reported in one study (Fernández Naveira et al., 2019) on syngas fermentation by *Clostridium carboxidivorans* in the presence of all trace elements during natural acidification of the medium from pH 6.2 to 5.0. The concentrations of acids decreased in successive experiments in the absence of selenium or tungsten by *Clostridium carboxidivorans* (Fernández Naveira et al., 2019).

The previous enrichment study providing the inoculum used in this work (Chakraborty et al., 2019) showed production of about 178 mg/L propionic acid though it quickly disappeared from the start of the solventogenic phase, indicating the presence of microbial populations producing uneven (C3) acids in the anaerobic inoculum. Hardly any previous study, on pure or mixed cultures, has detected the production of propionic acid from CO. In one study, it was observed that the addition of propionic acid to the medium can lead to the production of propanol with simultaneous syngas fermentation (supplemented with cornsteep liquor) in a mixed culture dominated by *Clostridium propionicum* and *Alkalibacterium bacchi* strain CP15 (Liu et al., 2014b). *Clostridium propionicum* is able to produce propionate and acetate via the acrylate-CoA pathway from substrates such as lactate (O Brien et al., 1990). The initial presence of lactic acid and propionic acid in the original inoculum (Chakraborty et al., 2019) indicates the possibility of the presence of microorganisms following this metabolic route, though further microbial analysis would be required to find the syntrophic association of CO/syngas metabolizing and propionate producing microorganisms.

5.4.1.3 Effect of lack of tungsten on the profile of CO consumption

In the absence of tungsten, from the start of the operation (Figure 1b), the effluent CO concentration was 1123.2 g/m^3 (inlet concentration being 1145.6 g/m^3) and it rapidly decreased to 350.35 g/m^3 on the 4th day of operation, corresponding to the highest gas consumption. During the 5th-10th days of operation, the effluent concentration was approximately 600-650 g/m^3. From the 11th to 15th day, the consumption again decreased and the outlet concentration was around 790 g/m^3. This period of operation, marking the solventogenic phase, also reveals the relative stability of the system with reference to the CO uptake by the bioreactor biomass. In the absence of tungsten, the production of CO_2, which was only occasionally measured, was found to be in correlation with the CO consumption. i.e., the rate of CO consumption, was proportional to the CO_2 production rate.

5.4.2 Mixed culture CO fermentation in absence of selenium (phase II)

5.4.2.1 Effect of lack of selenium on the production of metabolites and biomass in gas-fed bioreactors

After the above described experiment (phase I), the liquid medium was completely pumped out of the bioreactor while maintaining the sludge inside, and a new medium deficient in selenium was supplied to the bioreactor, though tungsten in the form of tungstate (2 mg/L) was now supplemented (phase II). Continuous feeding of CO was started again and a pH of 6.2 was maintained first to initiate acetogenesis. For the first 5 days of operation, the accumulation of acetic acid increased quite slowly and gradually to reach a concentration of 2.3 g/L, directly correlated with CO utilisation (Figure 2b). On the 6th day of operation, there was an abrupt rise in the concentration level yielding a maximum concentration of 6.6 g/L acetic acid in a short period (Figure 2a). Thus, the highest production rate of acetic acid, which was 4.3 g/L/d, was observed on the 5th day of operation. Butyric acid was also produced, although its production was more irregular. It is probable that, a few days later, once butanol started appearing, simultaneous butyric acid production and consumption took place as the amount of butyric acid detected in the medium remained low compared to the final concentration of butanol. It can be predicted that butanol was largely produced from the corresponding acids, i.e. butyric acid.

The operating pH was changed to 4.9 on day 6, and the alcohol concentrations, i.e. ethanol and butanol, started then suddenly to increase profusely. Ethanol production started levelling off slowly from day 14 to remain constant at its maximum value of 4.1 g/L between days 18-21 of operation, when the experiment was stopped. The butanol concentration kept increasing for a slightly longer period. In the last phase, on day 17, the highest amount of butanol was reached with a concentration of 1.88 g/L. As expected, no production of hexanol was observed in this case either, as no hexanoic acid had been detected in the acetogenic phase. Thus, similarly as with some pure cultures (Abubackar et al., 2015; Fernández-Naveira et al., 2019), the presence of tungsten stimulated the production of alcohols, i.e. ethanol and butanol in this study (Figure 1a), while its absence reduced the total concentration of alcohols (Figure 2a)

Selenium instead had no significant influence on the solventogenic stage. It was previously claimed by other authors (Saxena and Tanner, 2011) that in *Clostridium ragsdalei* cells, increasing the selenium concentration increased the production of ethanol from 38.32 mM (1.76 g/L) to 54.35 mM (2.5g/L) though it was reported not to have any effect of the growth rate of *Clostridium ragsdalei* in the presence of syngas as the carbon source (Saxena and Tanner, 2011). To the best of our knowledge, besides that paper, no other study has reported such positive effect of selenium on solventogenesis in any other *Clostridium* strain. Studies performed with pure cultures of *C. autoethanogenum* and *C. carboxidivorans* showed a clear positive effect of tungsten on solventogenesis, while such stimulating effect was not evident with selenium (Abubackar et al., 2015; Fernández-Naveira et al., 2019).

In the absence of selenium (Figure 2c), but in the presence of tungsten, the growth profile was gradual according to the OD recorded and increased from 0.04 to 0.53 from the start to the 4th day of operation. But between the 5th and 7th day of operation, the OD increased more sharply to 0.8 and afterwards 1.89, to later drop to 1.4 on the 11th day of operation and remain then stable till the end of the experiment. A clear logarithmic interpretation has been given in Figure S2. The μ_{max} value was found to be 0.64/d. Besides, it confirms that bacterial growth takes place during the acetogenic stage while no growth and even bacterial lysis occurred during the solventogenic stage.

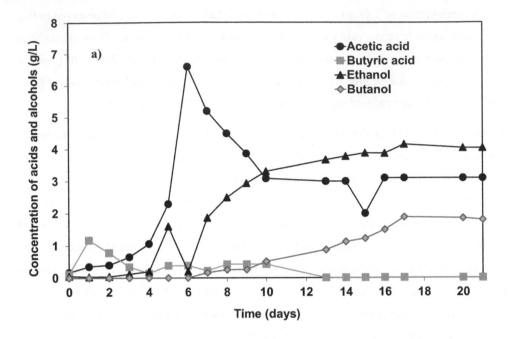

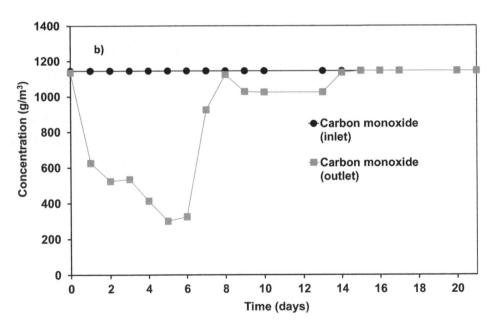

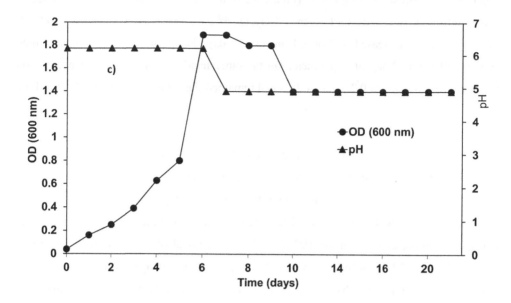

Figure 5.2: Continuous CO fed reactor experiments in the absence of selenium: (a) concentration of metabolites, (b) gas profile and (c) OD and pH

5.4.2.2 Effect of lack of selenium on the profile of CO consumption

The outlet CO concentration was $1135.2 \ g/m^3$ from the start of the operation (acetogenic phase) and with subsequent utilisation of CO, the outlet CO concentration steadily decreased to 300.6 g/m^3 on the 5th day of operation, but then abruptly rose to $924 \ g/m^3$ on the 7th day of operation. From the 8th to the 15th day of operation, the outlet concentration increased to $1145.6 \ g/m^3$ as no acetogenic fixation of CO was occurring. Both this and the previous experiment suggest that, irrespective of the presence or absence of selenium or tungsten, the highest amount of CO consumption takes place during the early stage of the bioreactor operation, concomitant with high biomass growth and the start of the acids production. After that, CO consumption decreases while the production of acids goes on and biomass growth levels off. Finally, CO consumption is at its lowest level when switching from acetogenesis to solventogenesis.

5.4.3 Batch assays with syngas/CO

Additional batch, bottles, assays, were performed, which allowed easy estimation of gas consumption data in such closed, constant volume systems and comparison with the bioreactor experiments. Syngas was used as substrate mixture and carbon and energy source in this case, and the results are compared with CO bioconversion data. In the absence of tungsten, and with syngas as the main substrate mixture, (Figure 5.3a), the acetic acid concentrations gradually increased from 0.1 g/L to 0.34 g/L on the 4th day. On 5th day, there was a sudden rise to 0.5 g/L acetic acid and it increased to 0.66 g/L on the 7th day. The concentrations remained then constant till the 10th day of experiment. On the other hand, the butyric acid concentration increased from 0.2 g/L to 0.3 g/L in 4 days and remained then stable at about 0.33 g/L till the 10th day. Though the ethanol concentration increased from 0.1 to 0.33 g/L, from the start till 3rd day of the experiment, it was readily consumed on the 4th day and no ethanol was found to be produced till the end of the experiment. The OD (600 nm) increased from 0.01 to 0.17 in five days, but remained unchanged later. On observing the gas consumption profile, the % removal of CO was found to increase from 0 to 41.9 after 8 days of operation and remained unchanged afterwards (Figure 5.3a.1). While there was a gradual increase in consumption of hydrogen reaching 28.6 % on the 10th day. The % removal of carbon dioxide fluctuated reaching 54% on 4th day of operation and 42.6 % on 10th day of operation; this decrease in removal was due to the fact that carbon dioxide can be produced through the consumption of carbon monoxide, meaning that at some stage more carbon dioxide was produced than originally present in the vials.

In absence of selenium and syngas as substrate (Figure 5.3b) the maximum concentrations of acids were similar (very slightly higher) as in the previous case, resulting in 0.8 g/L acetic acid and 0.46 g/L butyric acid, although the rate of acetogenesis was lower (in comparison to without W). A somewhat higher biomass concentration was reached, as indicated by an OD of 0.35 recorded on the 7th day of operation, when it then remained constant. In the course of 10 days of experiment, the acetic acid production rate increased from 0 to 0.13 g/L/day, till 4th day and decreased to 0.1 on 5th day of operation with a slight increase to 0.105 g/L /day on 7th day of operation. There was an abrupt fall in the acetic acid production to 0.01 g/L/day on 8th day, followed by slight consumption of about 0.01 g/L/day on the 9th day and finally the concentration of acetic acid became constant on 10th day of operation. On the other hand, butyric acid production rate increased from 0 to 0.1 on 1st day of operation and, followed by a small rise to 0.03 g/L/day on 2nd day of operation, remained constant at on 3rd and 4th day, and finally increased to 0.16 g/L/day on 5th day of operation. For the remaining course of the experiment, there were no more production of butyric acid. About 0.5 g/L of ethanol had accumulated on 5th day and its concentration then remained stable till the end of the experiment. The gas removal rates were higher (in comparison to without W) reaching maximum values of

116

64%, 79%, and 67.2% for carbon monoxide, carbon dioxide, and hydrogen, respectively, on the 10th day of operation (Figure 5.3b.1).

In the absence of both the trace elements, selenium and tungsten, the acids produced (Figure 5.3c), the gases consumed (Figure 5.3c.1) and the biomass growth were significantly lower compared to the previous set of results. No alcohols were produced at all (Figures 5.3c). They reached maximum concentrations of about 0.45 g/L acetic acid and 0.34 g/L butyric acid, when the biomass concentration OD reached a stable value. The rate of production of those acids were also lower than in the previous two experiments.

Pure CO is considered to be a better carbon and energy source for acetogens than syngas mixtures (CO, CO_2, and H_2). An additional experiment was performed in order to compare syngas and only CO as the substrates, and to confirm the inhibition of solventogenesis is in the absence of both trace elements with CO. Similarly as for syngas, only acids (acetic acid) and biomass growth was limited, with no production of alcohols (Figure 5.3d), confirming the effect of trace metals.

The lowest concentration of acids like acetic acid was 0.22 g/L and the initial amount of butyric acid which was about 0.2 g/L was totally consumed on the 1st day of operation.

(Figure 5.4d). The biomass growth was very limited and Butyric acid previously present in the medium was totally consumed. The OD reached 0.2 from a previously recorded data of 0.1 on the 1st day of operation and CO removal was only 12.5 % (Figure 5.4d.1).

The results obtained from the batch data indicates the cumulative role of selenium and tungsten on the production of acids and alcohols from gas fermentation. It is clearly observed that the metabolites production and biomass profile as well as the gas consumption are the lowest in the absence of both W and Se. The importance of tungsten+selenium> tungsten> selenium in the process of syngas fermentation in mixed culture to produce alcohols syngas can thus be inferred. When CO is the sole substrate, the production of acids, alcohols and biomass was observed to be the lowest as expected, due to the absence of H_2, (in comparison to syngas) which acts as the electron donor and facilitates the fermentation (Philips et al., 2017),

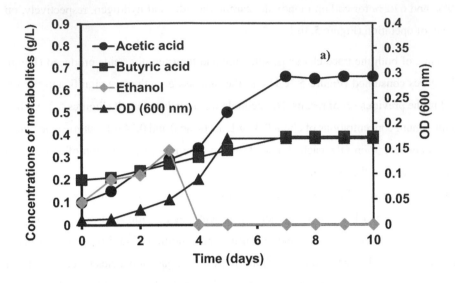

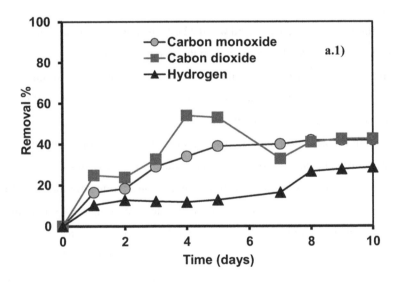

Figure 5.3: Batch experiments in absence of tungsten and syngas as substrate. a) Metabolites and biomass, a.1) Gas removal profile

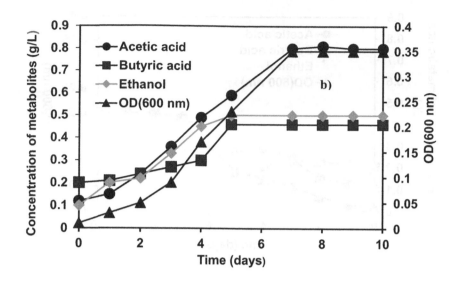

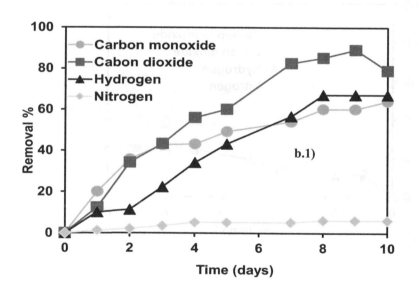

Figure 5.3: Batch experiments in absence of selenium and syngas as substrate. b) Metabolites and biomass, b.1) Gas removal profile

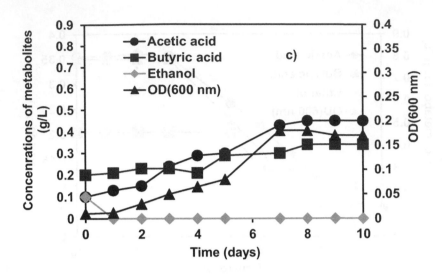

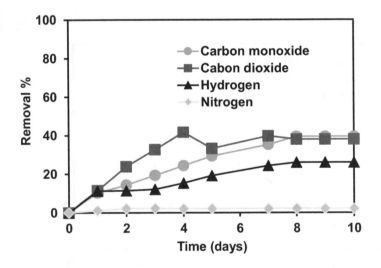

Figure 5.3: Batch experiments in absence of both selenium and tungsten and syngas as substrate. c) Metabolites and biomass, c.1) Gas removal profile

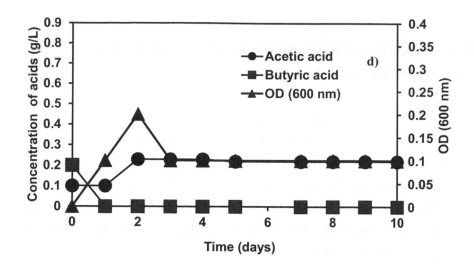

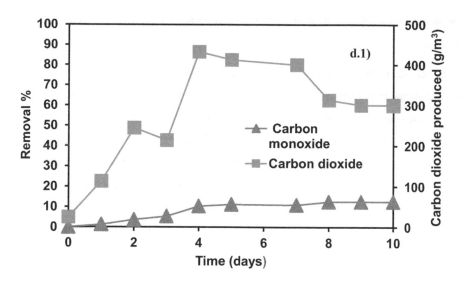

Figure 5.3: Batch experiments in absence of both selenium and tungsten and CO as substrate.
(d) Metabolites and biomass, (d.1) Gas removal profile

5.4.4 Predominance of tungsten over selenium as co-factor of solventogenic enzyme in the anaerobic sludge

Clostridial species contain selenium dependent FDH (formate dehydrogenase), but only NADP-dependent tungsten containing FDH from *C. thermoaceticum* has been isolated and characterized (Yamamoto et al., 1983). Some clostridial enzymes (with no detailed structural analysis) might require selenium and/or tungsten. The only predominant effect of selenium on acetogenesis and solventogenesis has been observed in *Clostridium ragsdalei* cells (Saxena and Tanner, 2011). Addition of selenium (in the form of selenate) to *Clostridium ragsdalei* cells increased the ethanol production from 38.3 (±2.12) mM (about 1.76 g/L) to 54.45 (±3.23) mM (2.508 g/L) and acetate production decreased from 9.56 (±1.34) mM (0.57 g/L) to 8.64 (±0.96 mM) (0.518 g/L). The average pH of the above mentioned study was 5.9 and experiments were done in batch bottles. The uncontrolled pH in that study might have had an additional effect on the strain´s behavior, besides the influence of trace metals. No study has been done to find thus for the effect of selenium or tungsten on the fermentation of CO by mixed cultures.

Chakraborty et al. (2019) described the enrichment of a solventogenic culture from methanogenic sludge in the presence of both selenium and tungsten. The acetic acid produced in the above mentioned study was completely converted to ethanol. In case of the absence of selenium and in the presence of tungsten, the acid production rates are higher compared to tungsten deficient operation, but there is a comparatively higher conversion of the accumulated acids (63%) to alcohols in absence of selenium. During the acetogenic phase, elimination of selenium led to the production of 6.6 g/L acetic acid (Figure 2a) which is the highest rate of acetogenesis compared to the tungsten deficient stage (7.3 g/L in 11 days) or the previously executed trace metal supplemented stage (6.2 g/L on the 10th day of acetogenesis or the 30th day of operation as observed by Chakraborty et al., 2019). It has to be considered that the sludge was previously accustomed to the presence of selenium (in the form of selenate) and tungsten (in the form of tungstate) (Chakraborty et al., 2019), and the absence of any of the trace elements may thus create a sudden shock for the enriched sludge. The experimental data thus showed the predominant effect of tungsten over selenium for solventogenic enzyme systems (Figure 1a).

5.4.5 Channelizing of metabolic flux in the two different stages of operation

The difference in the metabolic flux of acids and alcohols produced in the absence of tungsten and selenium was a noteworthy observation as found from the bioreactor experiments. In the absence of tungsten, butyric acid was converted to butanol during the solventogenic phase of CO fermentation (11th-15th day) as established by the consumption of 1.52 g/L butyric acid while 1.2 g/L butanol was produced. This observation indicates that the main flux of butanol production is via butyryl-CoA → butyrate → butaraldehyde (Fernández-Naveira et al., 2017), with an estimated gain of 0.4 mol ATP/mol of butanol (Bertsch and Muller 2015). The enzyme active in this route of butanol synthesis is aldehyde ferredoxin oxidoreductase (mediated by Fd^{2+}) or AOR. In the case of a pure culture of *Clostridium autoethanogenum*, excess tungsten induced AOR activity and solventogenesis (Abubackar et al., 2015). More research is required to find the effect of tungsten deficiency in mixed cultures.

In the absence of selenium and in the presence of tungsten, there was a sudden increase in butyric acid concentration to 1.17 g/L on the 2nd day of operation, but later a stable concentration of butyric acid of 0.42 g/L was observed, while the amount of butanol produced during solventogenesis was 1.8 g/L. Solventogenic conditions do not favour production of acids for the growth or survival of clostridial species. Since this highest detected concentration of butyric acid is not high enough to produce 1.8g/L butanol, there are two possible explanations. The metabolic flux for butanol production could be via direct conversion of butyryl-CoA to butanol rather than through the involvement of the AOR enzyme. However, the former pathway is less favorable from an energetic viewpoint than the latter involving AOR and conversion of butyric acid to butanol. The most reasonable scenario consists therefore in assuming that, in the presence of tungsten and absence of selenium, butyric acid was simultaneously produced and rapidly consumed to generate butanol.

The production of ethanol from acetic acid in both bioreactors (with or without either tungsten or selenium) was similar (Figure 1a and Figure 2a). Acetic acid was produced first, during the acetogenic stage, and ethanol started accumulating exactly once the acid reached its maximum concentration and started dropping in the solventogenic stage. This observation shows that the aldehyde ferredoxin oxidoreductase enzyme catalyzed the reduction of acetate to ethanol in the two phases of the bioreactor operation avoiding any ambiguity over its involvement compared to that of the butaraldehyde dehydrogenase. One major difference, in terms of bioconversion,

123

is that quite more acetic acid was converted to ethanol in the presence, rather than in the absence, of tungsten.

5.4.6 Inhibition of production of C6 metabolites

Hexanoic acid, and hence the corresponding alcohol, i.e. hexanol, was not produced in the absence of either selenium or tungsten as trace elements, indicating that the metabolic flux went mainly towards the synthesis of shorter chain fatty acids and alcohols. The enzymes responsible for production of hexanoic acid and hexanol are AAD (alcohol/aldehyde dehydrogenase) and ADH (alcohol dehydrogenase) (Fernández-Naveira et al., 2017) and no exhaustive research has been published on their activity and cofactor requirement. Further microbial analysis and deciphering the activity of C6 producing enzymes are required to reveal a similar role of specific trace metals, such as selenium and tungsten, in the production of C6 acids and alcohols. In pure cultures of *C. carboxidivorans*, production of 113 mg/L hexanoic acid at a pH of 6.2 was detected in the absence of tungsten (Fernández-Naveira et al., 2019). *C. carboxidivorans* is the only strain reported to produce hexanoic acid and hexanol through such a gas fermentation process. Perhaps, hexanol-producing strains were not present or may be lost or the mutual presence of both selenium and tungsten is required to significantly stimulate hexanol production in this enriched sludge.

5.4.7 Calculation of carbon balance

5.4.7.1 Calculation of carbon balance in phase I of tungsten deficient operation

CO is the sole carbon source and electron donor in this system of anaerobic fermentation. During the 11 days of acetogenesis, 183.7 g CO was supplied to the reactor containing 77.78 g C (carbon) according to calculations reported elsewhere (Chakraborty et al, 2019) considering the CO concentration to be 1145.6 g/m^3, and the gas flow rate 10 ml/minute. Theoretically, the amounts of CO that should have been consumed for acetic acid and butyric acid, according to the empirical stoichiometric equations (Fernández Naveira., et al 2017), are 13.362 g (5.724 g C) and 11.711 g (5.017g C), respectively, composing almost 13.64 % of the CO supplied. CO_2 produced is 10.72 g for acetic acid and approximately 11 g for butyric acid. As the solventogenic pH shift did not yield high amounts of alcohols and mostly alcohols are produced by acid reduction, only the acids have been considered here for calculation of the CO recovery. During acetogenesis, in the absence of tungsten, the cumulative CO_2 measured was 0.006 g/L, which is much less than the theoretical CO_2 to be produced. The calculations are based on the equations available elsewhere (Chakraborty et al. 2019). In the enrichment study (Chakraborty et al., 2019), as the final products were alcohols, the approximate CO recovered was calculated

124

to be 13.5 % of the supplied CO, which indicates the CO fixing capability of the original enriched microbial culture (Chakraborty et al., 2019). The biomass growth (as per OD from Figure 1c and figure 2c) has been unaccounted in this study as it is very low.

5.4.7.2 Calculation of carbon balance in phase II of selenium deficient operations

During the initial 6 days of operation, 100.2 g CO was passed through the medium. Theoretically, the CO consumption for the amount of acetic acid produced would be 12.313 g CO (5.277 g C) with 9.671 g CO_2 production, stating 12.3 % of the CO supplied has been recovered in the form of acetic acid. On the 6th day of operation, the recorded butyric acid concentration was 0.38 g/L. Using the empirical formula of butyric acid production from CO, 1.211 g/L CO has been used, covering approximately 1.20% of CO supplied.

In the second phase of operation, when the butyric acid is fully converted to butanol, yet there was excess production of butanol. The decrease of the acetic acid concentration is concomitant and proportional to the ethanol production. The excess butanol must be produced from CO supplied during this phase of operation or it could be that butyric acid production and assimilation occurs simultaneously. During the acetogenic phase, 1.17 g/L of butyric acid was recorded during the acetogenesis phase on the 2nd day of operation, which dropped to 0.78 g/L on the 3rd day of operation. As butyric acid was rapidly consumed in the acetogenic phase only and no concomitant production of butanol was observed, a stable concentration of 0.42 g/L butyric acid, even during the solventogenesis phase from 8th-10th day of operation.

When the pH was changed on the 6th day of operation, a rapid decrease in the butyric acid concentration was observed as 0.38 g/L of butyric acid decreased to 0.22 g/L of butyric acid with concomitant production of 0.16 g/L butanol. During the 8th-10th day of operation, the butanol production was found to increase from 0.253 g/L to 0.51g/L, giving a cumulative production of 0.257 g/L butanol. The butanol produced during this course of time is directly formed from CO and not directly from butyric acid as no fluctuation in butyric acid concentration was observed. Taking the empirical equation of butanol production from CO (Fernández-Naveira et al., 2017b), 1.486 g/L of CO supplied has been consumed. Now, 0.42 g/L of butyric acid was the stable concentration of butyric acid increased from 0.22 g/L butyric acid, yielding 0.2 (0.42-0.22g/L) g/L butyric acid during solventogenesis which was consumed afterwards. The theoretical amount of CO utilised is 0.69 g CO. Considering full conversion of butyric acid to butanol, 0.2 g/L of butyric acid would be converted to approximately 0.2 g/L butanol, which primarily comes from 0.69 g CO. Though 1.8 g/L butanol was produced, 0.16 g/L of butanol is converted from previously formed butyric acid, thus no CO was utilised in the solventogenic phase. From the remaining concentration of 1.64 g/L butanol, 0.2g/L butanol is produced from the intermediate butyric acid utilizing 0.69g/L CO. The rest of the butanol, i.e., 1.44 g/L butanol is produced directly from CO, i.e., about 8.32 g/L CO. In total, 8.94 g/L CO has been used. Alternatively, butyric acid could be simultaneously produced and converted to butanol and thus a mass balance calculation for the conversion of butyric acid to butanol is not possible for those specific metabolites. During the 15 days of reactor operation, 250.5g CO was supplied and taking 8.94 g CO, the approximate CO recovery is 3.56%. The total CO recovered is 17.06%, which is comparatively higher than the carbon recovered by the original inoculum (Chakraborty et al., 2019). The experimental (cumulative) CO_2 produced was 12.24 g/L CO_2. The calculations are based on the same equations as for tungsten deficient operation (Chakraborty et al. 2019).

5.4.8 DGGE analysis of the microbial community

DGGE analysis was performed on different samples to gain insights in the possible role of specific microorganisms in the gas fermentation process in the reactor biomass. *Clostridium autoethanogenum* is the only known CO/syngas fixing microorganism found in the absence of either tungsten or selenium in the bioreactor operation as observed from the DGGE analysis (Figure 4), though that organism was not detected in the inoculum (unacclimatized) or enriched sludge. The original isolate of *Clostridium autoethanogenum* from rabbit feces (Abrini et al. 1994) produces acetic acid and ethanol from CO/syngas mainly, and it was later found to produce 2,3 butanediol as well (Abubackar et al., 2015). The butanol and/or hexanol producing enriched sludge did not contain any known CO fixing strain reported to produce such alcohols. The only microorganism known from the literature to produce higher alcohols is *Clostridium carboxidovorans*. Further sequencing for individual microorganism's identification is required to further assign the taxonomy of the microorganisms present.

Some *Bacillus sp.* and *Clostridicae sp.* with more than 50% difference in genetic composition with *Clostridium autoethanogenum* were concurrently detected (Figure 5.4). It has to be mentioned here, that the used primer could only recognise bacterial sequences, and hence archaeal populations have not been determined in this study. As indicated in a previous study (Chakraborty et al., 2019b), methanogenesis was also observed during the enrichment of the anaerobic sludge at a low pH of 5.5 to obtain the solventogenic sludge used as inoculum in this study. A methanogenic population might thus be present in the bioreactor as well, although they are expected to be absent in the enriched sludge as no methane production was detected.

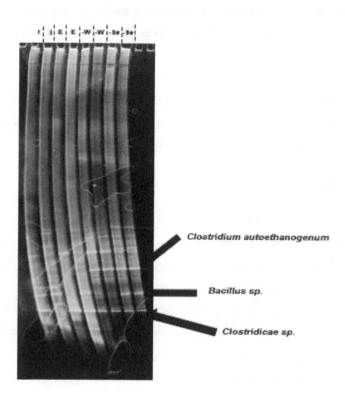

Figure 5.4: Microbial community analysis by DGGE. (I) Bands observed for Inoculum sludge, (E) Band observed for enriched sludge, (-W) Band observed for tungsten deficit culture, (-Se) Bands observed for selenium deficit cultures

5.5 Practical applications

Syngas is primarily composed of CO (and generally also H_2 and/or CO_2) and is emitted from varied sources like gasification of solid fuels like coal, petroleum coke, and biomass and can be a great feedstock for the production of biofuels (Abubackar et al., 2011). Lanzatech and Advanced Biofuels USA are two companies which are applying syngas/CO fermentation as a potential route for energy generation. Lanzatech is leading in syngas/CO fermentation for the production of alcohols and bioplastics (Ou et al., 2013). Besides, CO is the main compound found in some of the waste gases of steel producing industries. Yet, no study has been reported on the effect of trace elements on the bioconversion of CO or syngas in mixed cultures. Interestingly, mixed cultures do not require aseptic conditions and can tolerate oxidative stress. Trace element dosing has a significant impact on anaerobic mixed cultures and in industrial production of chemicals and products (Roussel et al., 2018). A varying dosing strategy for selenium or tungsten can modulate the production of alcohols at controlled pH conditions at larger scale. Omitting both trace elements can possibly restrict the CO conversion to volatile fatty acids and it indicates it has hardly any effect on the acetogenic capability of the microbial community. Further mass balance analysis and energy calculations would give a better idea about the cost efficiency and industrial feasibility of the process.

5.6 Conclusions

In this study, the presence of both selenium and tungsten were found to be vital for production of ethanol and butanol while tungsten (1.85 g/L ethanol and 1.2 g/L butanol were produced in the absence of tungsten) had a predominant effect over selenium (4 g/L of ethanol and 1.88 g/L butanol produced in absence of selenium) in the production of alcohols. Elimination of selenium has been found to increase acetogenesis, but was found to induce an abrupt increase of the growth rate of the CO fermenting microbial community. The mixed culture also exhibited a lower growth rate (0.55/d in the absence of tungsten) and (0.64/d in the absence of selenium) in comparison to the CO fermenting *Clostridium carboxidivorans* (2.4/d).

References

Abrini, J., Naveau, H., Nyns, E. J., 1994. *Clostridium autoethanogenum*, sp. nov., an anaerobic bacterium that produces ethanol from carbon monoxide. Arch. Microbiol. 161(4), 1994345-351.

Abubackar, H.N., Bengelsdorf, F.R., Dürre, P., Veiga, M.C., Kennes, C. Improved operating strategy for continuous fermentation of carbon monoxide to fuel-ethanol by *Clostridia*. Appl. Energy. 169 (2016): 210-217.

Abubackar, H.N., Veiga, M.C., Kennes, C. Biological conversion of carbon monoxide: rich syngas or waste gases to bioethanol. Biofuel. Bioprod. Biorefin. 5(2011): 93-114.

Abubackar, H.N., Veiga, M.C. and Kennes, C. Carbon monoxide fermentation to ethanol by *Clostridium autoethanogenum* in a bioreactor with no accumulation of acetic acid. Bioresour. Technol. 186 (2015) 122-127.

Abubackar, H.N., Fernández-Naveira, Á., Veiga, M.C., Kennes, C. Impact of cyclic pH shifts on carbon monoxide fermentation to ethanol by *Clostridium autoethanogenum*. Fuel, 178 (2016) 56-62.

Alves, J.I., Stams, A.J., Plugge, C.M., Alves, M., Sousa, D.Z. Enrichment of anaerobic syngas-converting bacteria from thermophilic bioreactor sludge. FEMS Microbiol. Ecol. 86 (2013): 590-597.

Bertsch, J., Müller, V., 2015. Bioenergetic constraints for conversion of syngas to biofuels in acetogenic bacteria. Biotechnol. Biofuels. 8 (2015): 210-231.

Chakraborty, S., Rene, E.R., Lens, P.N., Veiga, M.C. Kennes, C. Enrichment of a solventogenic anaerobic sludge converting carbon monoxide and syngas into acids and alcohols. Bioresour. Technol. 272 (2019): 130-136.

Fernández Naveira, Á., Veiga, M.C., Kennes, C.. HBE (hexanol butanol ethanol) fermentation for the production of higher alcohols from syngas/waste gas. J. Chem. Technol. Biotechnol. 92(2017), 712-731.

Fernández Naveira, Á., Veiga, M.C., Kennes, C. Glucose bioconversion profile in the syngas metabolizing species *Clostridium carboxidivorans*. Bioresour. Technol. 244 (2017): 552-559.

Fernández-Naveira, Á, Veiga, M.C., Kennes, C. Selective anaerobic fermentation of syngas into C2-C6 organic acids or ethanol and higher alcohols. Bioresour. Technol. 280 (2019): 387-395.

Fischer, C.R., Klein-Marcuschamer, D., Stephanopoulos, G. Selection and optimization of microbial hosts for biofuels production. Metab. Eng. 10 (2008): 295-304.

He, P., Han, W., Shao, L., Lü, F. One-step production of C6–C8 carboxylates by mixed culture solely grown on CO. Biotechnol. Biofuels. 11 (2018): 4-16.

Kennes, D., Abubackar, H.N., Diaz, M., Veiga, M.C., Kennes, C. Bioethanol production from biomass: carbohydrate vs syngas fermentation. J. Chem. Technol. Biotechnol. 91 (2016): 304-317.

Khanongnuch, R., Di Capua, F., Lakaniemi, A.M., Rene, E.R., Lens, P.N.L. H₂S removal and microbial community composition in an anoxic biotrickling filter under autotrophic and mixotrophic conditions. J. Hazard. Mat. 367 (2019) 397-406.

Köpke, M., Mihalcea, C., Bromley, J.C., Simpson, S.D. Fermentative production of ethanol from carbon monoxide. Curr. Opin. Biotechnol. 2 (2011): 320-325.

Kundiyana, D K., Raymond L. H., Wilkins, M.R. Effect of nutrient limitation and two-stage continuous fermentor design on productivities during *Clostridium ragsdalei* syngas fermentation. Bioresour. Technol. 102 (2011), 6058-6064.

Li, D., Meng, C., Wu, G., Xie, B., Han, Y., Guo, Y., Song, C., Gao, Z., Huang, Z. Effects of zinc on the production of alcohol by *Clostridium carboxidivorans P7* using model syngas. J. Ind. Microbiol. Biotechnol. 45 (2018): 61-69.

Liu, K., Atiyeh, H.K., Stevenson, B.S., Tanner, R.S., Wilkins, M.R., Huhnke, R.L., Mixed culture syngas fermentation and conversion of carboxylic acids into alcohols. Bioresour. Technol. 152 (2014a) 337-346.

Liu, K., Atiyeh, H.K., Stevenson, B.S., Tanner, R.S., Wilkins, M.R., Huhnke, R.L. Continuous syngas fermentation for the production of ethanol, n-propanol and n-butanol. Bioresour. Technol. 151 (2014b): 69-77.

Munasinghe, P.C., Khanal, S.K. Syngas fermentation to biofuel: evaluation of carbon monoxide mass transfer coefficient (k$_{La}$) in different reactor configurations. Biotechnol. Prog. 26(2010): 1616-1621.

O' Brien, D.J., Panzer, C.C., Eisele, W.P. Biological production of acrylic acid from cheese whey by resting cells of *Clostridium propionicum*. Biotechnol. Prog. 6 (1990): 237-242.

Ou, X., Zhang, X., Zhang, Q., Zhang, X., 2013. Life-cycle analysis of energy use and greenhouse gas emissions of gas-to-liquid fuel pathway from steel mill off-gas in China by the LanzaTech process. Front. Energ. 7(2013), pp.263-270.

Phillips, J., Raymond, H., Hasan, A. Syngas fermentation: a microbial conversion process of gaseous substrates to various products. Fermentation. 3 (2017), 28-53.

Roussel, J., Fermoso, F.G., Collins, G., Van Hullebusch, E., Esposito, G., Mucha, A.P. Trace element supplementation as a management tool for anaerobic digester operation: benefits and risks. IWA publishing London. (2018): pp-15

Saxena, J., Tanner, R.S., 2011. Effect of trace metals on ethanol production from synthesis gas by the ethanologenic acetogen, *Clostridium ragsdalei*. J. Ind. Microbiol. Biotechnol. 38 (2011): 513-521.

Shen, Y., Brown, R., Wen, Z. Syngas fermentation of *Clostridium carboxidivoran* P7 in a hollow fiber membrane biofilm reactor: Evaluating the mass transfer coefficient and ethanol production performance. Biochem. Eng. J. 85 (2014): 21-29.

Singla, A., Verma, D., Lal, B., Sarma, P.M. Enrichment and optimization of anaerobic bacterial mixed culture for conversion of syngas to ethanol. Bioresour. Technol. 172 (2014): 41-49.

Ungerman, A.J., Heindel, T.J. Carbon monoxide mass transfer for syngas fermentation in a stirred tank reactor with dual impeller configurations. Biotechnol. Prog. 23 (2007): 613-620.

Wan, N., Sathish, A., You, L., Tang, Y.J., Wen, Z. Deciphering *Clostridium* metabolism and its responses to bioreactor mass transfer during syngas fermentation. Sci. Rep. 7 (2017): 10090-10100.

Yamamoto, I., Saiki, T., Liu, S.M., Ljungdahl, L.G., 1983. Purification and properties of NADP-dependent formate dehydrogenase from *Clostridium thermoaceticum*, a tungsten-selenium-iron protein. J. Bio. Chem. 258 (1983): 826-1832. https://www.ncbi.nlm.nih.gov/pubmed/6822536.

Yan, S., Dong, D. Improvement of caproic acid production in a *Clostridium kluyveri* H068 and *Methanogen 166* co-culture fermentation system. AMB Expr. 8(2018): 175-187.

Yasin, M., Cha, M., Chang, I.S., Atiyeh, H.K., Munasinghe, P. and Khanal, S.K., 2019. Syngas Fermentation into Biofuels and Biochemicals. In Biofuels: Alternative Feedstocks and Conversion Processes for the Production of Liquid and Gaseous Biofuels. Academic Press. (2019): pp. 301-32

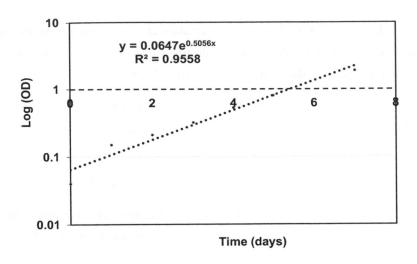

Figure 5.S1: Logarithmic growth of bacterial biomass in absence of tungsten

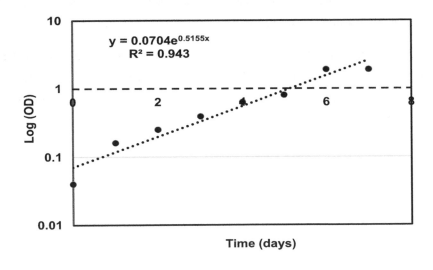

Figure 5.S2: Logarithmic growth of bacterial biomass in absence of selenium

Chapter 6

Recovery of alcohols by an in-house synthesized polymeric gel

This chapter is accepted as

Chakraborty S., Bera, R., Mandal A., Dey, A., Chakrabarty D, Rene, E.R., Lens P.N.L. Adsorptive removal of alcohols from aqueous solutions by N-tertiary-butylacrylamide (NtBA) and acrylic acid co-polymer gel. *Material. Today Communications*. 2019. (In press)

Abstract

A co-polymer gel of N-tert-butylacrylamide and acrylic acid was swelled in alcohols differing in its dipole moment and H-bonding interaction parameter so as to inspect the relative affinities of the gel towards alcohols. The gel swelled 900% in ethanol, 1200 % in propanol and 750% in butanol. The extent of swelling can be used as a guide for the preferential removal of particular alcohol from its mixture. Studies on adsorption isotherm using different isotherm equations helped in deciding the mechanism of adsorption of ethyl alcohol. The swelling and de-swelling kinetics of the gel in ethanol, propanol, and butanol (at low concentration of 5g/L each) followed mostly Fickian type diffusions for the processes of absorption and desorption. The scope of repeated use of the gel for alcohol recovery was investigated and every time approximately 98% alcohol was found to be recovered. The honeycomb morphology of the gel was accountable for such swelling of the gel.

Keywords: Swelling; de-swelling; alcohols; Fickian diffusion; morphology.

6.1. Introduction

Low molecular weight alcohols from sustainable biomass are considered as the most important alternative resources of fuels (Agarwal et al. 2007; Dürre et al. 2007). The recovery of alcohols (considered as biofuels) from fermentation broth is a big obstacle for full scale application because the recovery efficiency obtained by conventional methods is rather low. The conventional methods mostly in use today include adsorption (Hernandez-Martínez et al. 2018), pervaporation (Li et al., 2014), extraction (Qureshi et al, 1995), and gas stripping (Qureshi et al., 2001). All these methods suffer from disadvantages like high energy consumption, selectivity, high operating cost and fouling (pervaporation).

Absorption is a cost-effective and less energy intensive technique for recovering alcohols from fermentation broth (Abdallah et al., 2000), but the alcohols' neutral characteristics make it less liable for absorption, particularly in the presence of other polar organic compounds produced during the fermentation step (Abdallah et al., 1999). A highly selective adsorbent possessing high absorptivity (affinity) for specific alcohol is required for recovering the particular alcohol from the fermentation broth. Selectivity of the absorbing gel for a particular alcohol depends on the relative binding capacity of the adsorbent for recovering a particular alcohol versus water and hence the polarity, crystallinity and hydrophilicity of the alcohols play a great role (Oudshoorn et al., 2009). Desorption from the adsorbent is also an important issue in the recovery process. It has been noted that gradual heating might allow the stepwise desorption and, therefore, enable the recovery of the enriched alcohol (Milestone et al., 1981; Holtzapple et al., 1994).

An oraganogel is a class of gel composed of a liquid organic phase within a three dimensional, cross-linked network. The first step in the gel network formation is via polymerization which converts a precursor solution of monomers with various reactive sites (functional groups) into polymer chains that grow into a single covalently linked polymer network through cross linking. At a critical concentration (gel point), the polymer network becomes large enough so that on the macroscopic scale , the solution starts to exhibit gel like physical properties, an extensive continuous solid network with no steady flow and solid like rheological properties (Raghavan et al., 2012). However, organogels which are of "low molecular weight, gelators" can also be made to form gels via self-assembly. Secondary forces, such as Van der Waals or H bonding cause monomers to cluster into a non-covalent bonded networks that retain organic solvents and the networks grow and exhibit gel-like properties (Hirst et al., 2008). The uniqueness of polymeric organogels for such applications originates from its structure and the physico-chemical properties which can be engineered by the proper selection of monomers (comprising the gel), nature of cross-linkers and their doses, method of synthesis or use of chain regulators which ultimately determine its morphology. Organogels can be made to exhibit thermosensitivity, pH sensitivity (pH responsive smart gels), moisture and water sensitivity (smart hydrogel which adsorb large volume of water) (Samchenko et al., 2011) and sensitivity towards organic solvent like alcohols (organogels which adsorb and subsequently absorb large amounts of alcohol) (Hajighasen et al., 2013). Swelling of N-tert-butylacrylamide copolymers based hydrogels in water, ethanol, and DMSO mixtures, have been performed and temperature sensitivity of such gels were also tested (Ozturk et al., 2002; Ozmen et al., 2003). Organogels based on non-neutralized acrylic acid (60-94 mole %) and 2-acrylamido-2-methyl propane-sulphonic acid and also the co-polymers of acrylic acid and sodium styrene sulphonate are classical examples of super or high performance alcogels (Kabiri et al., 2011a). The swelled gels may be subsequently subjected to various treatments or conditions such that they release almost the entire amount of the entrapped solvent. Besides being used in engineering fuels, organogels find extensive applications for the controlled release of fragrance, drugs, and antigens (Kabiri et al., 2011a; Fadnavis et al., 1999; Kabiri et al., 2011b), surfactant industries, military applications (Bera et al., 2014), hand sanitizers, fire starters (Qingchun et al., 2007) and in enzyme immobilization (Kiritoshi and Ishihara 2002).

Kabiri et al. (2011c) prepared polymer organogels based on acrylic acid and sodium allyl sulphonate (SAS) by solution polymerization, using a persulphate initiator and a poly (ethylene glycol) diacrylate (PEGDA) crosslinker. This gel was treated with hydrochloric acid to remove the Na^+ counter ions which transform the gel from ionomer resin to a polyelectrolyte resin. The free radical polymerization was carried out to develop gels in presence of methanol which was used as solvent for copolymer synthesis.

This gel exhibited enhanced swelling, hence good absorption capacity for alcohols. In a separate report allied with the previous one (Kabiri et al., 2011c), Kabiri et al. (2013) converted water-absorbing polymer based on poly(sodium acrylate) to an alcohol absorbing polymer by the formation of an interpenetrating polymer network by immersing it in a solution containing 2-acrylamido-2-methylpropane sulphonic acid (AMPS), polyethylene glycol dimethacrylate and ammonium persulphate, the second network being formed in hydrated poly(sodium acrylate) through heating. The synthesized IPN (interpenetrating polymer network) gels have the ability to absorb up to 21 g of ethanol/g of gel and 39 g of methanol/g of gel, respectively (Kabiri et al., 2013). The co - polymer network of n-isopropyl acrylamide (NIPAAM) and 2-acrylomido-2-methylpropane sulphonic acid (AMPS) in the proportion of 3:97 can absorb ethanol up to 70g/g of gel. Co-polymers based on acrylamide and methacryl amidopropyltrimethyl ammonium chloride (MAPTAC) were synthesized in the presence of ammonium persulphate as initiator and polyethylene glycol diacrylate as the cross-linking agent. These acted as alcogels and were found to absorb up to 105.5 g of methanol/g of gel and 73.3 g of ethanol/g of gel, respectively (Hajihagsen et al., 2013).

Detailed gelling characteristics of N-tertiary-butyl acrylymide and acrylic acid co-polymer have been studied by Bera et al (21). N-tert-butylacrylamide (TBA) were found to be stable in solutions with ethyl acetate, ethanol, dichloromethane, water, formamide, N,N-dimethylformamide (DMF) and N,N-dimethylacetamide (DMAC) respectively at temperature ranging from 279.15 to 353.15 K (Gao et al., 2015). Such stability could be a good criteria for being a good temperature sensitive absorbent.

It is important to mention that several studies have already been done on the separation of short-chain alcohols by adsorption, mostly on silica aerogels and their organic derivatives (Wein et al., 2013), zeolites (Straathof et al., 2009; Silvestre et al., 2009), activated carbons (Qureshi et al., 2005) and polymer resins (Nielsen et al., 2009 a, b; Kabiri et al., 2011d), as adsorption requires less energy compared to other separation technologies. Studies on the separation of alcohols by polymeric/co-polymeric gels have however been sparsely reported.

A co-polymer of N-tert-butylacrylamide and acrylic acid (NtBA/AA) has been chosen such that the co-monomer (N-tert-butylacrylamide) disturbs the formation of a honeycomb structure solely formed by polyacrylic acid. Moreover, the immediate presence of the bulky structure of tertiary butyl groups (mainly hydrocarbon) decreases the hydrophilicity of the resulting gel to a larger extent without affecting the diffusion of the alcohol (less polar than water) inside the gel.

The objectives of the present work were as follows: (i) to inspect the absorptive capacities of the synthetic co-polymer gel NtBA/AA, in different linear aliphatic alcohols differing in its dipole moment and H-bonding action parameter, (ii) to study the batch kinetics of the sorption characteristic of the selected alcohols, (iii) to study the desorption characteristics of the gel swelled with different alcohols and, (iv) to determine the best adsorption isotherm approximating the characteristics of the particular gel-alcohol system.

6.2 Materials and methods

6.2.1 Materials, synthesis and consecutive studies

6.2.1.1 Materials

Acrylic acid, N, N′-methylene bisacrylamide (MBA) and azo bisisobutyronitrile (AIBN) were procured from Loba Chemie (Mumbai, India). N-tert-butylacrylamide (NTBA) was procured from Sigma Aldrich (Hamburg, Germany). All these chemicals were of analytical grade and were used as such without further purification.

6.2.1.2 Co-polymer synthesis

NtBA and acrylic acid were taken in equimolar proportions and were polymerized at 70°C in a two-necked round-bottomed flask using 0.1 mole% (of the total monomer) of AIBN as an initiator. Methanol was used as the solvent for polymerization. The flask was purged with a continuous stream of nitrogen (supplied from N_2 cylinders, Lyndell, India) in order to ensure an inert atmosphere which is an essential requirement for radical polymerization of acrylic monomers. N, N′-methylene-bisacrylamide was used as the crosslinker (0.01 mole %). The gelation took place after about 3 hrs. The gel was then taken out and soaked in methanol to remove any unreacted monomers. This procedure was repeated for ten times to ensure complete removal of monomers. The NtBA/AA co-polymer gel was then cut into small thin slices and allowed to dry first in the open air and then in a hot air oven (Annapurna India Ltd., India) maintained at a constant temperature at 100 °C for 2 h. The sample specimens were then cooled in a desiccator and weighed in previously weighed plastic containers with lid. It was again placed in the oven and the processes of drying and weighing were continued until a constant weight was achieved. The dried NtBA/AA gel samples were preserved in a vacuum desiccator (Basynth Pvt. Ltd., India). The method of synthesis closely resembled the previous work by Bera et al. [21].

6.2.1.3 Characterization

The NtBA/AA co-polymer gel was characterized with respect to its structural composition (spectroscopic) according to the protocol described by Bera et al. [21]. The gel samples were analyzed with respect to their structure and composition by Fourier Transform Infrared Spectroscopy (FTIR) and Nuclear Magnetic Resonance (NMR) spectroscopy.

6.2.2 Swelling studies

A known weight of the NtBA/AA co-polymer gel sample (0.1g) was allowed to swell in the desired solvent (ethanol, propanol, and 1-butanol with initial concentration of 5g /L) in an air tight container (plastic) at the controlled ambient temperature of 25°C. The samples were rectangular and disc shaped having the dimensions (2 cm x 2 cm x 1 cm). After 24 h, the sample was taken out from the solvent, wiped on the surfaces with blotting paper and transferred to an already weighed tightly closed plastic box. The sample was weighed and again transferred in the solvent for swelling. The processes of swelling and weighing were repeated once every 24h and the process continued until there was no further change in the weight of the swelled gel. Once this point was reached, the % swelling in that particular solvent was calculated (Eq. 1):

$$\% \text{ Equilibrium Swelling} = \frac{W_S - W_D}{W_D} \times 100 \qquad (1)$$

Where, W_S and W_D represent the weights of the swollen and dry gel in grams, respectively.

6.2.2.1 Estimation of solvent retention

For the purpose of calculating the percent deswelling and in turn the percent solvent retention at a given time during the process of swelling, the dried NtBA/AA gel sample of a given weight was allowed to swell in a particular solvent to reach its equilibrium capacity and the weight was taken. The swelled sample was then allowed to dry in open air and the weight of the dried sample was measured at definite time intervals. The percent solvent retention at a given time during the process of deswelling was estimated according to Eq. (2):

$$\% \text{ Solvent retention} = \frac{W_T - W_D}{W} \times 100 \qquad (2)$$

Where, W_T is the weight of the gel at any instant during deswelling, W is the weight of the gel sample with maximum solvent absorbed and W_D is the weight of the dry gel. All weights are in grams.

6.2.2.2 Swelling - deswelling kinetics

An absolutely dried gel sample piece was weighed accurately and kept immersed in the designated solvent kept in a 50 ml beaker and covered with a lid. The assembly is kept in a cabinet at the temperature of the laboratory. After 24 hours (exact time noted since immersing the sample in the solvent) the sample is taken out with the help of a tong and the solvent appearing on the surface of the sample is wiped out with the help of a piece of blotting paper and immediately transferred in an already weighed small air tight plastic container. The weight of the assembly was taken. The sample is then taken out carefully and put into the solvent again. After an equal interval of time the sample is again taken out, wiped and weighed following the same earlier method. This process was continued till the equilibrium is attained by the swelled samples. This weight is considered as M_∞. The earlier weights noted at different times was denoted as M_t for the sample tested at time t.

The swelling and deswelling data as obtained above were used to determine the corresponding kinetics according to Eq. (3):

$$F = \frac{M_t}{M_\infty} = k\,t^n \qquad\qquad (3)$$

Where, F is the dimensionless number representing the ratio of M_t, weight of swelled gel in grams at time t (h) to M_∞, the weight of the swelled gel in gram at equilibrium, k is the characteristic constant of the gel (s^{-1}) and n is the diffusional exponent (dimensionless) which indicates the transport phenomenon. From the slope of the plot, ln F vs ln t, the value of n, which indicates the type of diffusion governing the gel swelling, was calculated. Normally, when $n \leq 0.5$, Fickian diffusion is expected to take place and the diffusion rate is much slower than the relaxation rate. However, when $n = 1$, case II transport occurs and the process of diffusion is very fast with respect relaxation rate. An anomalous diffusion (Non-Fickian) occurs when the value of n lies between 0.5 and 1. The magnitude of exponent "n" and the nature of diffusion for various alcohols are given in Table 6.1.

Table 6.1: Type of diffusion with different values of n

Swelling			Deswelling	
Name of alcohols	Value of slope (n min^{-1})	Type of diffusion	Value of slope (n min^{-1})	Type of diffusion
Ethanol	0.41556	Fickian	0.64	Non-Fickian or anomalous
Propanol	0.41289	Fickian	0.48	Fickian
Butanol	0.21222	Fickian	0.22	Fickian

In order to investigate the scope of repeatedly using the gels for selective alcohol absorption, the NtBA/AA gel samples were prepared to undergo cyclical swelling and deswelling processes in succession separately for all the three alcohols (ethanol, propanol and 1-butanol) and their efficiencies were compared. In the present study, two cycles were carried out and in each step of swelling and deswelling, equilibrium was allowed to be reached. An increase in mass of the gel sample was determined at regular time intervals till the equilibrium was reached in each case. The swelling experiment was conducted at room temperature (25 °C) and at 1 atm. pressure. Equation (3) was used to determine the mode of swelling [Brannon-Peppas and Peppas 1989; Peppas and Bar-Howell 1986; Khare and Peppas 1995)

The deswelling of the ethanol, propanol or butanol loaded NtBA/AA gel samples was carried out by placing the swelled gel samples in the brine solution and the decrease in weight was noted at regular time intervals for each case until a constant weight of the de-swelled gel samples was achieved. Thereafter, the data were fitted in the plot of ln (M_t/M_∞) vs ln t in the same manner as that was followed earlier for determining the swelling kinetics.

6.2.2.3 Adsorption isotherm

Alcohol solutions were prepared using sterile deionized water at a range of initial concentrations (1, 2, 3, 4 and 5) g/L. The equilibrium adsorption experiments were performed in sterile 17ml, glass Hungate tubes containing 100mg of the gel sample and the remainder aqueous alcoholic solutions prepared. The tubes were completely filled such that no head-space was available promoting maximal contact between the gel sample and the liquid. Sealed tubes were used to allow the headspace gas to be completely vented through a needle as the aqueous solution was introduced via a second needle and syringe. In this way the samples were not pressurized. Samples were equilibrated while being agitated at 150rpm on an orbital shaker at 300rpm for 48 hours. The specific loading of alcohol adsorbed to the gel at equilibrium was determined by the following mass balance relationship (Eq. 4).

$$Q_e = (C_{aq0} - C_{aqe}) V_{aq} /m \qquad (4)$$

Where C_{aq0} and C_{aqe} were initial and the equilibrated aqueous alcohol concentrations, respectively. V_{aq} is aqueous phase volume and m is the mass of the gel. All adsorption experiments were performed in triplicate.

In order to understand the interactions between the adsorbate (ethanol, propanol and butanol in the present case) and the adsorbent (NtBA/AA co-polymer gel), approximately 0.1g of gel sample was allowed to swell in 1-5 g/L of ethanol (aqueous solution) in 100 mL air-tight plastic cups. In each case, observations were continued until swelling reached its equilibrium. The data of adsorption thus obtained for each alcoholic concentration were fitted to the three most extensively used linear equilibrium adsorption isotherm models, namely Freundlich, Langmuir, and Temkin. The best fitting curve was chosen as the appropriate model governing the interaction between the adsorbate and the adsorbent at the surface of the adsorbent. The Freundlich isotherm is represented by Eq. (5):

$$Q_e = K_f C_e^{1/n} \qquad (5)$$

Where, Q_e is the amount of alcohol adsorbed (mg/g) at equilibrium and C_e is the adsorbate equilibrium concentration (mg/L); K_f (mg/g) and n (intensity of adsorption) are the Freundlich constants (signifying the nature of isotherm) and the intensity of adsorption in the particular case of adsorbate and adsorbent (gel). On taking the logarithm of Eq. 5, it stands:

$$\ln Q_e = \ln K_f + \frac{1}{n} \ln C_e \qquad (6)$$

On plotting $\ln Q_e$ against $\ln C_e$, a straight line is obtained from Eq. (6). The constant values can be calculated from the intercept and the slope of the plot.

The Langmuir isotherm equation is represented by:

$$\frac{C_e}{Q_e} = \frac{1}{b} Q_o + \frac{C_e}{Q_o} \qquad (7)$$

The equation (7) can be rewritten as

$$\frac{1}{Q_e} = \frac{1}{b} \frac{Q_o}{C_e} + \frac{1}{Q_o} \qquad (8)$$

Where, Q_o (mg/g) and b (L/mg) are the Langmuir constants related to the adsorption capacity and energy of adsorption, respectively. The constants were calculated from the plot of $1/Q_e$ vs $1/C_e$.

The Temkin's isotherm equation is represented by:

$$Q_e = \frac{RT \ln A}{b} + \frac{RT \ln C_e}{b} \qquad (9)$$

Where, Q_e (mg/g), the amount of alcohol adsorbed at equilibrium is plotted against $\ln C_e$ the concentration of the adsorbate at equilibrium. R stands for the universal gas constants and T represents the temperature in Kelvin at which the adsorption process was carried out. From the intercept and the slope of the plot, the values of, respectively, RT ln A/b and RT/b are obtained. All these 3 equations are in the linear form of y=mx+c and thus linear plots are generated. In this case, an attempt has been made to find out the best fitting linear curve which explains the adsorption of the alcohol on the adsorbate surface explicitly. The Langmuir, the Freundlich and the Temkin isotherm might adequately describe the same set of liquid-solid adsorption data in a certain concentration ranges, in particular, if the concentration is small and the adsorption capacity of the solid gel is large enough. All three isotherm equations approach a linear form.

6.2.3 Analytical methods

6.2.3.1 Measurement of alcohols

Aqueous alcohol concentrations were measured using a high performance liquid chromatograph (HPLC,1100 series, Agilent Santa Clara, CA), fitted with a AZORBAX eclipse XDB-C18 column (Agilent Santa Clara, CA). A mobile phase of deionized water at a constant flow rate of 1.0 mL/min was used. A refractive index detector was used for detecting the analytes. External standards were used for determining the alcohol concentrations. 1 mL of sample was collected and diluted with deionized water for analysis in the HPLC. The elution times for ethanol, propanol and 1-butanol were 11.56, 23.5 and 46.8 min, respectively.

6.2.3.2 Morphology: Optical microscopy

The surface morphology of the NtBA/AA gel in dry and wet (swelled) conditions was inspected by phase contrast microscopy using a KRUSS microscope (Kruss Optronic, Germany) at a magnification of 100 ×.

6.2.3.3 Scanning electron microscopy

Scanning electron micrographs (SEM) of the fractured surfaces of both the dry and swelled NtBA/AA gels were taken using a scanning electron microscope (S-3400N Hitachi JAPAN). The samples were gold coated with coating time of 45 s using a gold coater Hitachi E 1010. CRYO–SEM was performed at -160 °C after fracturing the swollen NtBA/AA gel sample at -

187 °C followed by gold coating with 2 min coating time in an argon atmosphere to analyze the pore structure using a SEM (Carl Zeiss EVO18, Dusseldorf, Germany).

6.3 Results and Discussions

The relatively hydrophobic monomer N-tert-butylacrylamide was co-polymerized with the hydrophilic monomer acrylic acid to optimize the hydrophilic-hydrophobic properties of the ultimate cross-linked co-polymer gel network. The approximate chemical structure of the resulting co-polymer network is shown in Figure 6.1. Both intra- and inter-molecular H-bonding between the CO−NH group of N-tert-butylacrylamide and the−COOH group of acrylic acid moieties leads to the formation of a large number of sites for solvent absorption besides the previously formed sites by the cross-linked network.

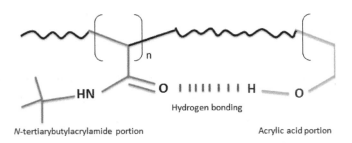

Figure 6.1: Hydrogen bonding between the monomers of the NtBA/AA copolymer gel.

The secondary valence bonds as obtained by hydrogen bonding act as pseudo-crosslinks leading to the formation of tiny enclosed empty domains which act as sites for solvent absorption. The ratio (1:1) of the two co-monomers was selected after several observations with different ratios of the same co-monomers. The selection was made based on maximum percent swelling exhibited by different co-polymer gels with different co-monomer ratios [Bera et al., 2014]. In such a co-polymer gel, the swelling behavior in different solvents is controlled besides the physical parameters like temperature, pressure and concentration difference, not only by the steric effects of the bulky tert-butyl group, but also by the network structure formed by the interactions of the reactive sites of the two participating monomers of N-tert-butylacrylamide and acrylic acid.

6.3.2 Influence of dipole moment on swelling characteristics of the gels

Figure 6.2 shows the mode of changes in the swelling behavior of the NtBA/AA gel as a function of the dipole moment of different alcoholic swelling media (methanol being the most polar and -butanol the least polar of the range of alcohols considered). It is apparent that the percent equilibrium swelling of the NtBA/AA gel in alcoholic media undergoes a steady and steep increase at relatively low dipole moments in the range studied. This enhancement in swelling was observed within a very short range. Once maximum swelling occurred in a solvent with dipole moment at around 1.68 D, a reduction in equilibrium swelling was observed in the swelling pattern with further increase in dipole moment of the swelling medium. This indicates the non-ionic characteristics of the NtBA/AA gel. Therefore, it can be expected that these gels are unable to undergo sufficient dissociation into ions, particularly in the case of relatively polar solvents like methanol and water to produce the necessary osmotic pressure difference that is usually responsible for swelling (Weihn et al., 2013, Straathof et al., 2009).

Alcohol absorbency is dependent on characteristics such as hydrogen bonding ability of the solvent with the gel, −OH/C (hydroxyl to carbon) ratio, electronic features such as dipole moment, dielectric constant and steric hindrance of neighboring groups of the solvent −OH groups (Kabiri et al., 2011b). It was confirmed from Figure 6.2 that the absorbency and hence the percent swelling gradually increases as the dipole moment decreases in going from methanol to propanol, while in the case of 1-butanol the −OH/C ratio decreases to such a great extent that its absorbency in the NtBA/AA gel falls abruptly. This observation also follows the principle of dependency of swelling behavior of the NtBA/AA gel on its ionic and non-ionic characteristics (Kabiri et al., 2011c; Brannon-Peppas and Peppas 1989).

Considering the series of aliphatic alcohols, methanol having the shortest hydrocarbon chain length is supposed to have less polarity than water, but the highest one amongst the alcohols under study. The basic objective of carrying out swelling test in methanol along with the other solvents with the NtBA/AA gel is to have an idea of the polarity of the gel (in turn its ionic character) and hence to predict its sorption characteristic. From this test, it is apparent that its polarity and hence hydrophilicity of the NtBA/AA gel is appreciably less than methanol (which is completely miscible with water in all proportion). The hydrophilicity, is however, more than 1-butanol. This test shows the extent to which the tailoring of hydrophilic–hydrophobic character of the NtBA/AA gel has been achieved with the incorporation of NtBA into the structure of polyacrylic acid (PAA). Thus, the polarity of the gel matches the range between methanol and 1-butanol, a property intensely desired in the present study concerning the recovery of ethanol from its mixture with other alcohols.

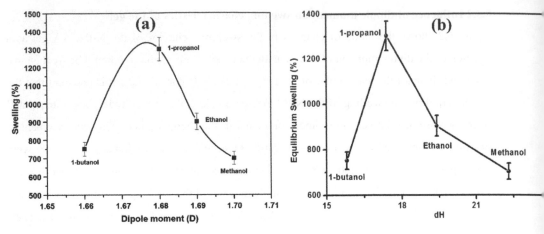

Figure 6.2: Percent swelling of the NtBA/AA gel versus the dipole moment of the adsorbate

6.3.3. Kinetics of gel swelling in different alcoholic media

From Figure 6.3, which exhibits the subject plots for the three solvents investigated, namely ethanol, propanol and 1-butanol, the values of 'n' (the ratio of $\Delta y/\Delta x$) were determined as 0.41556 for ethanol, 0.41289 for propanol and 0.21222 for 1-butanol. All the values of 'n' were < 0.5 and follow predominantly the Fickian type of diffusion. This diffusion process is considered to be a slow one in relation to the process of relaxation. In the inset of Figure 3, a comparative study of such 'n' values for the three different alcohols has been made using bar charts. Although propanol has an almost identical value of 'n' approaching that of ethanol, 1-butanol has a much slower diffusion rate than expected. It was observed that the value of 'n' decreases as the hydrocarbon chain length of the alcohol (swelling medium) increases, i.e. as the dipole moment decreases. This indicates that the alcohol transport mechanism becomes relatively more Fickian as the ionization (and hence the osmotic pressure) of the gel increases, causing the diffusion in the medium gradually to decrease. The dynamic swelling behavior of a cross-linked polymer gel is known to be dependent upon the relative magnitude of diffusion and polymer relaxation time (Peppas and Bar-Howell 1986). It is well known that on entry of solvent into the structure of the gel due to diffusion, the network structure of the gel gets disturbed and if the polymer chains do not get sufficient time to relax and accommodate it, further adsorption of solvent molecules on the gel surface cannot take place. Thus, it is an essential requirement that for incessant dynamic diffusion, the relaxation rates of the polymer chains should be slightly faster than the diffusion rate.

Among the three alcohols tested, the relative proneness of the swelling media to undergo ionization follows the order ethanol > propanol > 1-butanol and the diffusivities rates in the NtBA/AA gel followed the same order. The swelling behavior of any polymer network is expected to depend on the nature of the polymer forming the gel, the polymer-solvent compatibility and the degree of cross-linking. The polymer elasticity combined with the polymer solvent mixing contributes to the overall swelling process (Khare and Peppas 1995; Subrahmanyam et al., 2015). The osmotic pressure causing the diffusion controlled kinetics in the present case is given by (Ren et al., 2011):

$$\pi \text{ (operating osmotic pressure)} = \pi_{mixing} + \pi_{elastic} + \pi_{ion} \qquad (10)$$

Where, the LHS of the equation denotes the effective osmotic pressure causing diffusion in the gel and the RHS gives the algebraic sum of the contributions of the constituent factors contributing to this.

In the present case, $\pi_{elastic}$ remained the same in the three cases of the solvents (characteristic response of the polymer towards osmosis). The π_{mixing} (compatibility), the measure of affinity of the NtBA/AA gel towards a particular solvent and its contribution towards osmosis as well as π_{ion}, a measure of the degree of ionization of the gel in particular solvent, controlled the osmotic pressure. The polymer-solvent compatibility is the highest in case of ethanol (solubility parameters being very close besides their dipole characteristics). The scope of ionization is the highest in case of ethanol due to its relatively higher polarity amongst the solvents. Thus, the gels swelled to 900% in ethanol, 1200% in propanol and 750% in 1-butanol.

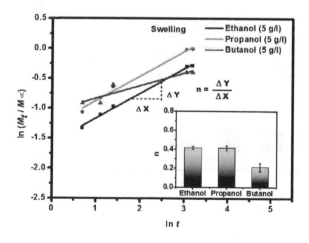

Figure 6.3: Logarithmic plot of M_t/M_∞ versus swelling time of the NtBA/AA gel for different alcohol

6.3.4 Kinetics of deswelling

As expected, an identical pattern of deswelling curves (Figure 4) resembling those obtained during swelling were observed. The NtBA/AA gel samples swelled in different alcoholic solvents show values of the slope 'n' > 0.5 for ethanol during deswelling, while the samples swelled in the other two alcohols exhibit 'n' values < 0.5. Thus, the deswelling processes of the gels swelled in propanol and 1-butanol likely still follow a Fickian diffusion process during deswelling. However, the deswelling process in case of the gel sample swelled in ethanol follows a non-Fickian or anomalous diffusion. The reason may be attributed to the fact that once entrapped within the cellular structure of the gel, the alcohol concentration being higher, the small ethanol molecules can undergo association through hydrogen bonding within itself (intra molecular hydrogen bonding through the H of one molecule with the more electronegative O atom of the OH group of another ethanol molecule, this proneness to undergo H bonding decreases with increase in hydrocarbon chain length) or with the amide linkages of the NtBA/AA gel material. Thus, it does not exist in the simple discrete molecular form which could have undergone Fickian diffusion. This possibility is remote in case of either propanol or 1-butanol, which are less polar because of the higher hydrocarbon chain length and rather inert towards the types of secondary reactions as in the case of ethanol.

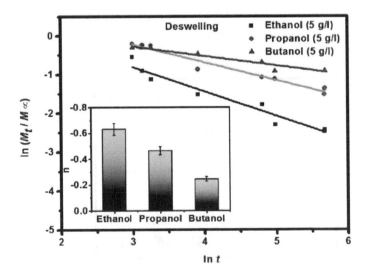

Figure 6.4: Logarithmic plot of M_t/M_∞ versus deswelling time for different alcohols from the NtBA/AA gel.

6.3.5 Cyclical swelling and deswelling

Figure 6.5 shows the mode of cyclical absorption and desorption for the three alcohols under investigation. The co-polymer gel is the most efficient in absorbing ethanol compared to the other two alcohols. In case of each solvent (ethanol, propanol and 1-butanol), a slight decrease in efficiency in the second cycle was observed. In addition, the rate of absorption reduces remarkably in the second cycle. The amount of solvent retained at the end of each cycle remains almost the same in the three cases. For each alcohol, the extent of swelling reaches its equilibrium (different for each alcohols) at an equal time interval and de-swells to its maximum (equilibrium) at the same time interval. This signifies a very good elastic response of the gel towards tensile elongation (cage expansion) and tensile contraction (cage retraction) of the gel. This characteristics can be attributed to the well-known enhanced elastic recovery (rheological characteristics) of the co-polymer which predominates over its viscous component (Dey et al., 2014; Patachia et al., 2007)

The process of co-polymerization, also known as the process of internal plasticization, usually reduces the glass transition temperature (T_g) of a co-polymer (Chen et al., 2000) The T_g of the synthesized co-polymer is usually intermediate between the two homo polymers formed by the constituent monomers. This co-polymerization increases the rubbery behavior, which in turn helps it to undergo large deformation at small stress. From the mode of solvent uptake during swelling in each cycle, it is evident that the entry of solvent into the cellular domain of the gels (as exhibited in the morphology, Figure 6.7) is quite fast (burst entry) and then slows down while nearing to its equilibrium value. During deswelling, a similar burst release of solvent followed by a slow and sustained release near its equilibrium stage can occur. However, the rate of burst entry and burst release decreases in their corresponding rates during the successive cycles. The phenomenon of incomplete deswelling (less than 100 percent release of solvent) may possibly be attributed to the chemisorptive forces exerted by the hydrogen bonding of the solvents (through the H of the H-O-R group) on the core of the cell wall due to the groups' accessibility of the $CO-NH_2$ group of acrylamide (NTBA). The chemisorption force gradually diminishes after the surface of the cells gets covered by a layer of alcohol. In the case of ethanol, the chemisorption force was observed to be the highest (Figure 6.5).

The greater the chain length of the hydrocarbon portion of the alcohol (which makes it relatively non-polar), the slower is the diffusion rate within the same time interval. A similar observation was made by Khare and Peppas (1995), who concluded that the gels did not lose their elasticity during the cyclical use.

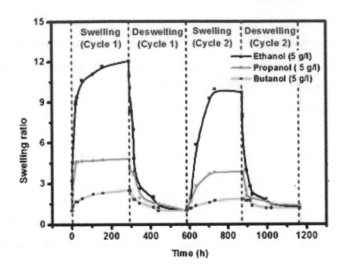

Figure 6.5: Cyclical swelling deswelling of the NtBA/AA gel in ethanol, propanol and 1-butanol.

6.3.6 Adsorption isotherm

From Figure 6.6(a), the values of the different parameters are given in Table 6.2 for the different isotherms of Freundlich, Langmuir and Temkin. An essential feature of the Langmuir isotherm is expressed by a dimensionless constant called separation factor R_L, also called equilibrium parameter, which is defined as $R_L = 1/1 + bC_0$ where C_0 (mg/L) is the initial adsorbate concentration. The value of R_L indicates the shape of the isotherm to be either unfavourable ($R_L > 1$), linear ($R_L = 1$), favourable ($0 < R_L < 1$) or irreversible $R_L = 0$. In the present case, R_L equals 0.999, which lies in the range $0 < R_L < 1$ (Ghosh et al., 1990). Thus, the adsorption was favourable.

An estimate of the adsorption strength, i.e. the attractive force between the adsorbent and the adsorbate, can be made by estimating the Gibbs' free energy change of the adsorption (ΔG_0) for the alcohol under study. In case of the Langmuir adsorption isotherm, the Gibbs free energy change is given as $\Delta G_0 = -RT \ln b$ (Nethaji et al., 2013; Kyuo et al., 2008)], where b stands for the Langmuir constant (0.1150 L/g) or 5.405 L/mol ($R = 8.314 \times 10^{-3}$ kJ/mol/K). Here, the value of $\Delta G_0 = -4.18$ kJ/mol for the Langmuir isotherm. This observation indicates that the adsorption can be readily reversed, a necessity for adsorbent regeneration and ultimate recovery of the alcohol (Weihn et al., 2013) As the negative ΔG_0 is not too high (i.e., not too much negative), a strong interaction such as chemisorption does not appear to take place at the interface, i.e. a weak interaction as the Van der Waals interaction is operative at the interface. The magnitude of 1/n in the Freundlich isotherm quantifies the favourability of adsorption and degree of heterogeneity on the adsorbent gel surface. A value of 1/n < 1 as in the present case suggests the favourability of adsorption. From the Temkin's plot, the term RT/b is a constant and related to the heat of adsorption and the term A stands for equilibrium binding constant.

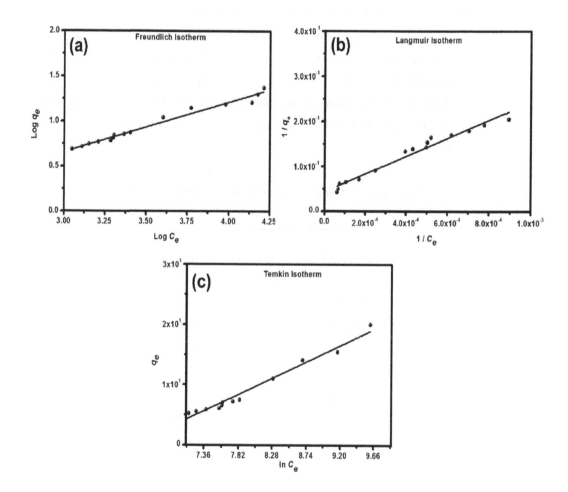

Figure 6.6: (a) Freundlich isotherm: plot of log q_e versus log of equilibrium concentration C_e, (b) Langmuir isotherm: plot of 1/q_e vs1/C_e and (c) Temkin isotherm.

The three plots as formulated for the three different isotherms following the respective equations are represented in Figure 6.6 Statistical regression analysis was carried out to determine the value of the Pearson's 'r' coefficient for each case in order to find out the deviation (inversely the closeness) of the actual values from the theoretical linearity in each individual case. It is important to mention here that the closer the value of 'r' to unity, the more is the appropriateness of the equation applied. For cross-verification of the results, the 'Chi-square' values were also calculated from the statistical analysis.

On comparing the statistical parameters obtained for the three isotherm equations, the Pearson's 'r' values (data not shown) for the three cases were close to unity. However, the 'Chi-square' value in case of the Langmuir isotherm was much lower (9.8×10^{-19}) than those obtained from the other two cases, (2.2×10^{14}) for Temkin and 0.00126 for the Freundlich isotherms. Thus, the Langmuir adsorption isotherm explains the mode of interaction between the adsorbate and the adsorbent in a well-defined manner in comparison to the other two modes of adsorption.

Table 6.2: Values of Freundlich, Langmuir and Tempkin constants

Freundlich	K_f (l/kg)	0.3688
	n (g/l)	1.8097
Langmuir	Q_0 (l/g)	22.8415
	b (mg/g)	0.1150
Tempkin	A (l/g)	0.0017
	b	422.1095

6.3.7 Mechanism of swelling

The overall swelling of the NtBA/AA gel occurred in two steps. Due to the presence of some polar groups, namely amides and acids ($-CONH_2$ and $-COOH$), the alcohol molecules get attracted to them (polar-polar interaction and H-bonding). These molecules adhere to the surface through adsorption by weak Van der Waals forces. The adsorbed layer of molecules being loosely held (as indicated by the value of ΔG_0' the Gibbs' free energy change due to adsorption) start diffusing into the core of the cell formed by the network. Thus, a combination of adsorption and absorption leads to the swelling phenomenon. Surface adsorption is a step prior to the bulk process of absorption as explained by Deyko and Jones (2012).

6.3.8. Morphology: Optical microscopy

The microscopic images of the NtBA/AA gels in dry and swelled conditions are given in Figure 6.7. The closely knitted small microscopic cells in the dry gel undergo an increase in hydrodynamic volume after absorbing solvent to its maximum possible extent as dictated by its equilibrium swelling values, which is mostly dependent on the capacity of the NtBA/AA gel material to undergo tensile deformation closely associated with its rheological parameters, tensile viscosity and cell wall thickness. The uniqueness of the cell structure is that, most of the cells remain intact and discrete even after undergoing deformation (Bera et al., 2014]. This can be considered as supportive evidence of the elastic nature of the co-polymer constituting the partition wall of the cell which entraps the solvent. This elastic nature as mentioned previously in "cyclical swelling and deswelling" helps in expanding when absorption takes place and contracting when desorption occurs.

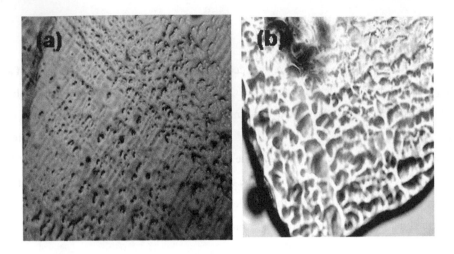

Figure 6.7: Microscopic image of the NtBA/AA gel: (a) dry condition and (b) swelled condition.

6.3.9 Scanning electron microscopy

The SEM images of both the dry and swelled NtBA/AA gel surfaces are shown in Figure 6.8. It is important to mention that the CRYO-SEM technique was employed in the case of swelled gels to make the irregular cellular regions more distinct and clear. Figure 8 shows that the dry gel surface is full of microscopic empty cells (mostly available in the dark regions) which on absorbing alcohol increase in dimension in the swelled state. However, these changes in dimensions do not occur uniformly in all directions because of statistical variations in the wall thickness. Here, the expanded cells are not symmetrical. The cells appear to be more or less discrete in nature. The extent of swelling, and hence the amount of solvent absorbed, depends on the stretching or tensile deformation characteristics of the cell wall. In order to envisage the mode of swelling, the SEM images of the gel in two different magnifications were considered. It is interesting to note that the dry gel contains a large number of phase-separated domains of different sizes and shapes. These domains most probably develop because of the inhomogeneous distribution of entangled polymer chains which form a strong association within itself due to the combined effect of intra and intermolecular hydrogen bonding. Micro-heterogeneity of the gel surface enhances the pore sizes in gels (Bera et al., 2014).

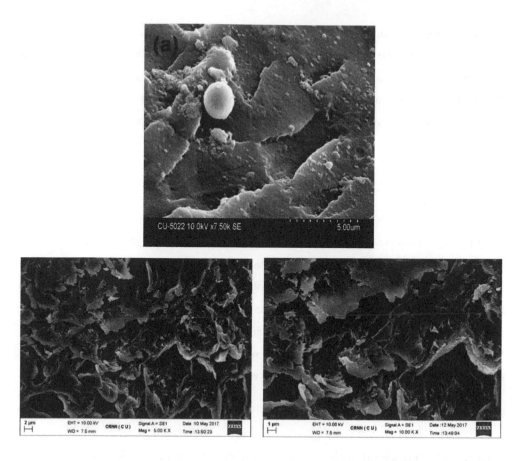

Figure 6.8: SEM images of the NtBA/AA copolymer gel: (a) dry, 100× (b) swelled in ethanol, 100× and (c) cryo-SEM image of the swelled gel at 500×.

6.4 Practical applications of the synthesized gel

In previous studies (Weihn et al., 2013; Straathof et al., 2009; Qureshi et al., 2005), the most commercially utilized adsorbents for alcohols were found to be silica, zeolite and their organic derivatives. Qureshi et al. (2005) observed that the amount of energy required for recovering 1-butanol was 1.948 kcal/kg of 1-butanol and the adsorptive capacity is 206-252 mg/g adsorbent. Another adsorbent used for adsorbing ethanol is modified activated carbon AC400 (Silvestre et al., 2009) with an adsorptive capacity of 9.76 g/100g of adsorbent. In comparison to the above two cases, the synthesized NtBA/AA gel in the present study requires no extra energy because the experiments were carried out under static conditions. About 1.9 g/l of 1-

butanol and 5 g/l of ethanol could be adsorbed by 0.1 g of dry gel. In a recent study, Seo et al. (2018) demonstrated the applicability of molecular sieving carbon (MSC5A) to separate gas-phase ethanol with an adsorptive capacity of 0.613 g/g MSC5A. However, converting the dissolved ethanol to gaseous ethanol required excess energy and hence the process is expensive. Thus, the synthesized NtBA/AA gel is able to overcome the obstacles of the previously reported adsorbents, though further research is necessary for full scale industrial applications.

6.5 Conclusions

A primarily non-ionic co-polymer gel consisting of equimolar proportions of N-tert-butylacrylamide and acrylic acid was found to be effective in absorbing alcohols from an aqueous mixture even at substantially low alcohol concentrations. The dry gel was found to have a reticulum or network of tiny microscopic cells of non-uniform size and dimension. The Fickian mode of diffusion allowed the solvent molecules to imbibe into the interior of the cells. The gel adsorbed mostly the low molecular weight of primary alcohols, which could be completely recovered and the gel showed good reversibility characteristics.

References

Agarwal, A.K. Biofuels (alcohols and biodiesel) applications as fuels for internal combustion engines. Prog. Energy. Combust. Sci. 33 (2007): 233-271.

Abdallah, D.J., Lu, L., Weiss, R.G. Thermoreversible organogels from alkane gelators with one heteroatom. Chem. Mat. 11 (1999): 2907-2911.

Abdallah, D.J., Weiss, R.G. Organogels and low molecular mass organic gelators. Ad. Mater. 12 (2000): 1237-1247.

Bera, R., Dey, A., Chakrabarty, D. Studies on Gelling Characteristics of N-Tertiary Butyl Acrylamide–Acrylic Acid Copolymer. Adv. Polym. Technol. 33 (2014) 21387.

Brannon-Peppas, L., Peppas, N.A. Solute and penetrant diffusion in swellable polymers. IX. The mechanisms of drug release from pH-sensitive swelling-controlled systems. J. Control. Release 8 (1989) 267-274.

Chen, J., Shen, J. Swelling behaviors of polyacrylate superabsorbent in the mixtures of water and hydrophilic solvents. J. Appl. Polym. Sci. 75 (2000): 1331-1338.

Dada, A.O., Olalekan, A.P., Olatunya, A.M., Dada, O. Langmuir, Freundlich, Temkin and Dubinin–Radushkevich isotherms studies of equilibrium sorption of $Zn2+$ unto phosphoric acid modified rice husk. IOSR J. Appl. Chem. 3 (2012): 38-45.

Deyko, A., Jones, R.G. Adsorption, absorption and desorption of gases at liquid surfaces: water on [C 8 C 1 Im][BF 4] and [C 2 C 1 Im][Tf 2 N]. Faraday discussions. 154(2012): 265-288.

Dey, A., Bera, B., Bera, R., Chakrabarty, D. Influence of diethylene glycol as a porogen in a glyoxal crosslinked polyvinyl alcohol hydrogel. RSC Adv. 4 (2014) 42260-42270.

Dürre, P. Biobutanol: an attractive biofuel. Biotechnol. J.: Healthcare Nutrition Technology 2 (2007) 1525:1534.

Fadnavis, N.W., Koteshwar, K. An Unusual Reversible Sol– Gel Transition Phenomenon in Organogels and Its Application for Enzyme Immobilization in Gelatin Membranes, Biotechnol. Prog. 15 (1999) 98-104.

Gao, X., Xue, W.L., Zeng, Z.X., Fan, X.R. Determination and Correlation of Solubility of N-tert-butylacrylamide in seven different solvents at temperatures between (279.15 and 353.15 K). J. Chem Eng. Data, 60 (2015) 2273-2279.

Ghosh, P. Polymer Science & Technology of Plastics and Rubbers, Tata Mcgaw hills publishing company Limited, New Delhi, 1990.

Hajighasem, A., Kabiri, K. Cationic highly alcohol-swellable gels: synthesis and characterization. J. Polym. Res. 20 (2013), 218.

Hernandez-Martinez, A.R., Lujan-Montelongo, J.A., Silva-Cuevas, C., Mota-Morales, J.D., Cortez-Valadez, M., De Jesus Ruíz-Baltazar, Á., Cruz M., Herrera-Ordonez. Swelling and methylene blue adsorption of poly (N, N-dimethylacrylamide-co-2-hydroxyethyl methacrylate) hydrogel, React. Funct. Polym. 122 (2018): 75-84.

Hirst, A.R., Coates, I.A., Bouchetean, T.R., Miravel, J.F., Escuder, B. Castelleto, V., Hamley, I.W., Smit, D.K. Low-molecular-weight gelators: elucidating the principles of gelation based on gelator solubility and a cooperative self-assembly model. J. Am. Chem. Soc. 130 (2008): 9113-9121.

Holtzapple, M.T., Brown, R.F. Conceptual design for a process to recover volatile solutes from aqueous solutions using silicalite. Sep. Technol. 4 (1994): 213-229.

Kabiri., K, Azizi, A., Zohuriaan-Mehr, M.J., Marandi, G.B., Bouhendi, H., Jamshidi, A. Super-alcogels based on 2-acrylamido-2-methylpropane sulphonic acid and poly (ethylene glycol) macromer, Iran. Polym. J. 20 (2011a) 175-183.

Kabiri, K., Lashani, S, Zohuriaan-Mehr, M.J., Kheirabadi, M. Super alcohol-absorbent gels of sulfonic acid-contained poly (acrylic acid). J. Polym. Res. 18 (2011b) 449-458.

Kabiri, K., Azizi, A., Zohuriaan-Mehr, M.J., Bagheri M. G., Bouhenoi, H. Alcohophilic gels: polymeric organogels composing carboxylic and sulfonic acid groups. J. Appl. Polym. Sci. 120 (2011c) 3350-3356.

Kabiri, K., Roshanfekr, S. Converting water absorbent polymer to alcohol absorbent polymer. Polym. Adv. Technol. 24 (2013) 28-33.

Khare, A.R., Peppas, N.A. Swelling/deswelling of anionic copolymer gels. Biomaterials 16 (1995) 559-567.

Kiritoshi, Y., K. Ishihara, K. Preparation of cross-linked biocompatible poly(2-methacryloyloxyethyl phosphorylcholine) gel and its strange swelling behavior in water/ethanol mixture. J. Biomater. Sci. Polym. Ed. 13 (2002) 213-224.

Kuo, C.Y., Wu, C.H., Wu, J.Y. Adsorption of direct dyes from aqueous solutions by carbon nanotubes: Determination of equilibrium, kinetics and thermodynamics parameters. J. Coll. Interface Sci. 327 (2008): 308-315.

Li, Y., Wee, L.H., Martens, J.A., Vankelecom, I.F. ZIF-71 as a potential filler to prepare pervaporation membranes for bio-alcohol recovery. J. Mat. Chem. A 2 (2014): 10034-10040.

Milestone, N.B., Bibby, D.M. Concentration of alcohols by adsorption on silicalite. J. Chem. Technol. Biotechnol. 31 (1981): 732-736.

N.A. Peppas, B.D. Bar-Howell, Preparation methods and structure of hydrogels. In Hydrogels in Med and Pharm, (1986), 1-26.

Nethaji, S., Sivasamy, A., Mandal, A.B. Adsorption isotherms, kinetics and mechanism for the adsorption of cationic and anionic dyes onto carbonaceous particles prepared from Juglans regia shell biomass. Int. J. Env. Sci. Technol. 10 (2013): 231-242.

Nielsen, D.R., Prather, K.J. In situ product recovery of n-butanol using polymeric resins. Biotechnol. Bioeng. 102 (2009a) 811-821.

Nielsen, D.R., Amarasiriwardena, G.S., Prather, K.J. Predicting the adsorption of second generation biofuels by polymeric resins with applications for in situ product recovery (ISPR). Bioresour. Technol. 101 (2009b) 2762-2769.

Oudshoorn, A., Vander Wielen, L A., Straathof, A.J. Assessment of options for selective 1-butanol recovery from aqueous solution. Ind. Eng. Chem. Res. 48 (2009) 7325-7336.

Ozmen, M.M., Okay, O. Swelling behavior of strong polyelectrolyte poly (N-t-butyl acrylamide-co-acrylamide) hydrogels. Eur. Polym. J. 39, (2003) 877-886.

Ozturk, V., Okay, O. Temperature sensitive poly (N- t-butyl acrylamide-co-acrylamide) hydrogels: synthesis and swelling behavior. Polymer 43, (2002) 5017-5026.

Patachia, S., Valente, A.J., Baciu, C. Effect of non-associated electrolyte solutions on the behaviour of poly (vinyl alcohol)-based hydrogels. European Polym. J. 43 (2007) 460-467.

Qureshi, N., Maddox, I.S. Continuous production of acetone-butanol-ethanol using immobilized cells of *Clostridium acetobutylicum* and integration with product removal by liquid-liquid extraction. J. Ferment. Bioeng. 80 (1995): 185-189.

Qureshi, N., Blasehek, H.P. Recovery of butanol from fermentation broth by gas stripping. Renew. Energ. 22 (2001): 557-564.

Qureshi, N., Hughes, S., Maddox, I.S., Cotta, M.A. Energy-efficient recovery of butanol from model solutions and fermentation broth by adsorption. Bioproc. Biosyst. Eng. 27 (2005) 215-222.

Qingchun, Z., Changling, Z. Synthesis and characterization superabsorbent-ethanol polyacrylic acid gels. J. Appl. Polym. Sci. 105 (2007) 3458-3461.

Raghavan, S.R., Douglas, J.F. The conundrum of gel formation by molecular nanofibers, wormlike micelles, and filamentous proteins: gelation without cross-links? Soft Mater, 8 (2012): 8539-8546.

Ren, H.Y., Zhu, M., Haraguchi, K. Characteristic swelling–deswelling of polymer/clay nanocomposite gels. Macromolecules 44 (2011) 8516-8526.

Samchenko, Y., Ulberg, Z., Korotych O. Multipurpose smart hydrogel systems. Adv. Colloid Interface Sci. 168(2011): 247-262.

Seo, D.J., Takenaka, A., Fujita, H., Mochidzuki, K., Sakoda, A. Practical considerations for a simple ethanol concentration from a fermentation broth via a single adsorptive process using molecular-sieving carbon. Renew. Energ. 118 (2018): 257-264.

Silvestre, A.J., Silvestre, A.A., Sepulveda, E.A., Rodriguez-Reoriso, F. Ethanol removal using activated carbon: Effect of porous structure and surface chemistry. Micropor. Mesopor. Mat. 120 (2009) 62-68.

Straathof, A.J., Oudshoorn, A., Vander Wielen, I.A.M. Adsorption equilibria of bio-based butanol solutions using zeolite. Biochemical engineering journal, 48 (2009) 99-103.

Weihn, M., Levario, T.J., Staggs, K., Linnen, N., Yuchen, W., Pffer, R., Lin, Y.S., Nielsen,

D.R. Adsorption of short-chain alcohols by hydrophobic silica aerogels. Ind. Eng. Chem. Res. 52 (2013) 18379-18385.

Chapter 7

Outlook, conclusions and perspectives

7. General discussion

7.1. Introduction

The increasing scarcity of non-renewable sources of energy has propelled various interest in finding research opportunities to exploit the enormous quantities of wastes generated by various industries to produce value added chemicals along with the simultaneous detoxification of the waste. The great impetus now being felt in this domain of research has enabled the development of various state-of-the-art bioremediation techniques for the detoxification of gaseous pollutants such as CO/CO_2 and liquid phase pollutants such as phenol and oxyanions of metalloids. Although various challenges and hurdles have been taken on the way to reach the goals, there are also some promising success stories.

In this thesis, various biological routes were developed for the detoxification of polluted effluents from different sources, including liquid effluents and gaseous effluents from oil refinery and other allied industries. Petrochemical waste is a big concern in both the developed and developing countries, starting from Bangladesh (Azad et al., 2015) to USA (Psomopoulos et al., 2009). This dissertation is a documentation of the bioconversion of phenolic compounds and selenium ions which are toxic and obnoxious constituents in the liquid effluents of oil refineries and detoxification and bioconversion of CO/syngas (gaseous effluents of oil refineries) to value added chemicals (acids and alcohols)

In the second chapter, initially, a concise description of the presence of phenol and selenium in the effluent streams and measures taken to remove them have been described. The process parameters of CO/syngas fermentation have been reviewed in Chapter 2 of this thesis. The influence of physical and biochemical parameters, namely the operating pH, temperature, enzymes and the metallic co-factors related to solventogenic enzymes producing alcohols associated with the syngas fermenting microorganisms have been described in this chapter. In the last 50 years, the amount of research carried out in this field is relatively low due to the toxicity of CO and difficulty in maintaining the strict anaerobic conditions required under laboratory conditions.

In this dissertation, while investigating the simultaneous removal of phenol and selenite ions, there was an endeavor to investigate whether biogenic processes could influence the reduction of selenium oxyanions, frequently present in wastewater in the form of selenite and selenate, leading to the production of nano Se(0). Keeping in mind the extensive applications of biogenic nano Se(0) in the areas of food, steel, cosmetic, glass and energy (Mal et al., 2016), a biogenic process could be engineered to produce nano-sized Se in an appreciable quantity.

Biofuels from CO and syngas are gradually becoming "the most popular future fuels" after the extensive debates on using lignocellulosic fuels (Mayfield and Wong, 2011) and considering the gradual depletion of petro based fuels (Campbell 2002). Thus, the following key issues were addressed in this thesis: (i) aerobic detoxification of phenolic effluents with the simultaneous reduction of selenite to Se(0) by a co-culture of *Phanerochaete chrysosporium* and *Delftia lacustris* (Chapter 3), (ii) enriching methanogenic anaerobic sludge and determination of the influence of selenium and tungsten deficiency for the production of C_2-C_6 carboxylates from CO and syngas (Chapters 4 and 5), and (iii) separation and simultaneous purification of the desired biofuels (alcohols) from the fermentation broth using an in-house synthesized polymeric gel (Chapter 6).

A co-polymeric resin has been put to test for such separation of the bio-ethanol from its mixture with various other alcohols which are simultaneously produced. This topic has been the subject matter of Chapter 6. Chapter 7 provides an insight into the current scenario and future perspectives of the different avenues of research in the concerned fields.

7.2 Detoxification and valorisation of liquid wastes from the petroleum refinery

Effluents of a petroleum refinery contain different types of toxic phenolic compounds including phenol, cresols (*o*-cresol, *m*-cresol and *p*-cresol), nitrophenols (2-nitrophenol, 4-nitrophenol, chlorophenols, 2,4-dichlorophenol, 2,6-dichlorophenol, 2,4,6-trichlorophenol, and 2,4,5-trichlorophenol) (Salcedo et al., 2019) and selenite ions. Chapter 3 exemplifies the syntrophic association of the fungus *Phanerochaete chrysosporium* and bacterium *Delftia lacustris* to degrade phenol and simultaneously detoxify selenite to industrially useful Se(0) (Chakraborty et al., 2019a). The fungus *Phanerochaete chrysosporium* has the ability to detoxify phenol and to reduce selenite ions to nano Se(0) (Werkeneh et al., 2017). In this study, the degradative capability of phenol by the fungus *P. chrysosporium* was found to increase substantially in the presence of *Delftia lacustris*, while the selenite reducing capacity by the co-culture was also found to be enhanced compared to those obtained by the pure cultures of *P. chrysosporium*. During the detoxification process, although nano Se(0) was produced as expected, further research is necessary to decipher the interlinking of the metabolic pathway of the fungus and the bacteria to produce Se(0).

7.3 Valorisation of gaseous waste streams from the oil refinery

Chapter 4 demonstrates the bioconversion of the gaseous wastes namely CO, CO_2 and H_2 to useful fuels by enriching methanogenic (CH_4 producing) sludge to a solventogenic (alcohol producing) sludge. The influence of pH, addition of L-cysteine hydrochloride and yeast extract were evaluated in a 2L bioreactor (Chakraborty et al., 2019b). Hexanol (1.46 g/L), ethanol (11.1 g/L) and butanol (1.8 g/L) were produced from CO after ~40 d of reactor operation.

7.4 Importance of deficiency of tungsten (W) and selenium (Se) in CO fermentation

Based on the results of Chapter 4, the selective production of acids and alcohols was monitored in the absence of two trace metals, either tungsten (W) or selenium (Se). This is the first report that shows the effect of trace elements on mixed culture CO conversion. In a previous study, increasing the concentration of Se from 1.06 mM to 5.03 mM in the medium containing *Clostridium ragsdaleii* showed an increase in the production of ethanol from 1.6 g/L to 2.5 g/L in 4 days (Saxena and Tanner, 2011). In pure cultures of *Clostridium carboxidivorans* (Fernández-Naveira et al., 2019), the absence of W negatively affected the production of alcohols, while the excess addition of 0.75 µM W (in the form of tungstate) resulted in no accumulation of acetic acid (Abubacker et al., 2015). In this study, the absence of W yielded ~ 1.8 g/L of ethanol (pH 4.9), while the amount of acetic acid produced was ~ 7.34 g/L. A similar operation with two consecutive periods of pH maintained at 6.2 first and subsequently decreased to 4.9, yielded 6.6 g/L of acetic acid at high pH, and 4 g/L of ethanol as well as 1.88 g/L butanol at the lower pH in Se deficient medium.

7.5 Novel application of a polymeric gel for the adsorptive recovery of alcohols

The process of bioconversion of CO/syngas is usually accompanied with the production of a mixture of alcohols. Therefore, it is necessary to recover the alcohols from the fermentation broth using a suitable adsorbent. Chapter 6 focusses on the recovery and separation of the solvents produced from CO fermentation. A co-polymeric gel, synthesized in house for the removal of alcohols, was shown to be an affordable and effective route to recover alcohols from the fermentation broth (Chakraborty et al., 2019c). The conventional method of fractional distillation for the recovery of alcohols uses considerable amounts of energy. On the other hand, commercially available resins are rather expensive and the amount of resin required is usually high. The synthesized co-polymeric gel showed a very high efficiency to recover alcohols, ~15 times its own weight. In comparison to the commercial resins such as L-493 and Dowex Optipore to recover ethanol and butanol (Sadrimajd et al., 2019), the co-polymeric gel was found to be durable and more effective in recovering the alcohols.

The schematic representation of the overview of results and future perspectives is given in the following page in Figure 7.1

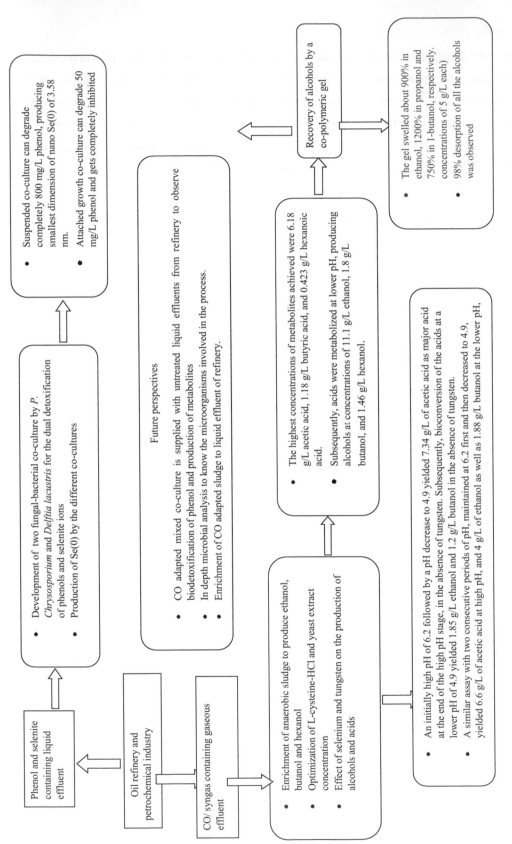

Figure 7.1 Overview of thesis results and future perspectives

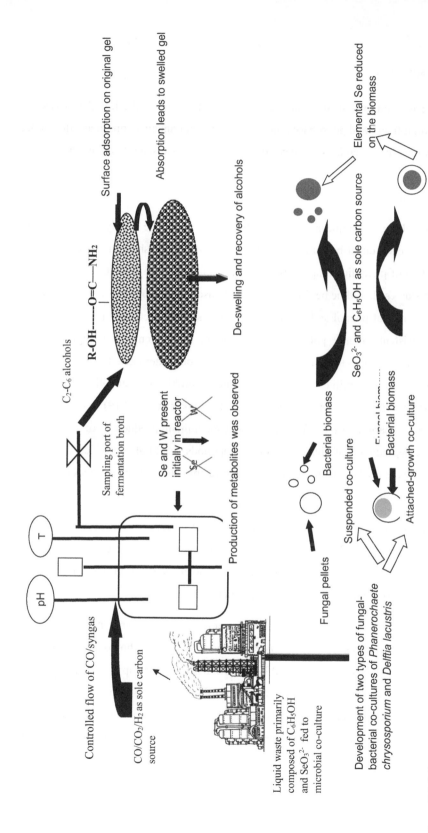

Figure 7.2 Schematic showing the bioconversion of gaseous and liquid effluents of oil refinery and petrochemical industry to value-added chemicals and their recovery

7.6 Future perspectives

In this PhD thesis, a biological route has been followed to detoxify the liquid and gaseous effluents from oil refinery and petro-based industries and simultaneously produce value-added chemicals. However, commercial scale detoxification of the toxic selenite compounds, along with the simultaneous degradation of phenolic compounds and production of nano Se(0) is yet to be achieved in practice.

The aerobic co-culture technique proposed in this thesis would be highly effective at the pilot/industrial scale as fungal-bacterial syntrophic metabolism was shown to be effective for the biodegradation of phenolic compounds and reduction of selenite ions (Chapter 3). More detailed kinetic studies should be performed in order to ascertain the biomass growth rate and the detoxification rate of phenol and selenite ions. As this study used only phenol, the most simple toxic persistent organic pollutant (POP), future experiments with higher carbon containing phenols like p-cresol and nitrophenol should be undertaken to understand the degradative capability and limiting factors of this unique aerobic co-metabolism between *D. lacustris* and *P. chrysosporium*. In a recent study, Wadgaonkar et al. (2019) has reported the selenate reducing capability of *D. lacustris* with concomitant production of seleno-ester compounds. The degradation of higher phenolic compounds would also produce an array of interesting organic metabolites which would give a clear profiling of the metabolic pathway involved in its metabolism.

Large quantities of Se oxyanions are discharged from the mining industries and coal-fired power plants (Ellwood et al., 2016), along with phenols and polyphenolic compounds. Thus, apart from waste streams discharged from oil refineries, the effluents of mining and coal-fired power plants could also be treated using the proposed technique.

The structure and composition of the biogenic Se(0) produced, i.e. at different nano-dimensions using the different co-cultures of *P. chrysosporium* and *D. lacustris*, should also be analyzed by X-ray photoelectron spectroscopy (XPS) and Raman spectroscopy. The extracellular polymeric substance (EPS), as visible from transmission electron microscopy (TEM) observations, could be further characterized using fluorescence excitation emission matrix (FEEM) spectroscopy, to investigate the humic and protein compounds associated with nano Se(0) (Sheng and Yu, 2006)

Focusing on the detoxification and valorizations of the gaseous emissions from oil-refineries via CO/syngas fermentation is a niche area and ever growing in the field of bioenergy. However, there are only few publications that focus on the capability of mixed cultures to perform CO/syngas fermentation. In this thesis, experiments were carried out for a total period of 46 d in a bioreactor, wherein the first 21 d were used for enriching the anaerobic methanogenic sludge to acetogenic/solventogenic sludge (Chapter 4). After sludge enrichment, the biomass could be utilized in a pilot-scale bioreactor for production of higher volatile fatty acids and alcohols. The concentration of L-cysteine-HCl and yeast extract was varied to observe their effect on the productivity. Future experiments should be conducted using cheaper supplements like corn steep liquor (Liu et al., 2014) for inducing the biomass growth rate, acetogenesis and solventogenesis steps. A primary aspect of future experiments would be using the liquid effluent of an oil refinery for medium supplemented with CO fermenting anaerobic microbes.

Oil refinery effluents contain a lot of trace elements like selenite which is an important trace element in syngas fermentation. The effect of the absence of one trace element, i.e. either W or Se, on the performance of the enriched sludge was tested to estimate their cumulative effect on CO/syngas fermentation (Chapter 5). Besides, the functional microorganisms responsible for CO/syngas fermentation were determined by DGGE (denatured gradient gel electrophoresis). However, for a more in-depth understanding of the microbial communities involved, Illumina sequencing should be done to identify the shift in microbial community of the CO/syngas fermenting microorganisms (Elnaker et al., 2018).

It is noteworthy to mention that the phenolic compounds can induce toxicity to the species of solventogens e.g., *Clostridium beijarincki* (Liu et al., 2018), though directly no experiments have been performed to find the toxicity of phenol on CO fermenting solventogens. In such a case, simultaneous enrichment of the microbes present in the liquid effluent and the anaerobic solventogens could produce important volatile fatty acids and gases like CH_4 (phenol can be converted to CH_4), (Young and Rivera 1985).

The development of a novel co-polymeric gel able to recover alcohols is a pivotal outcome of this PhD thesis. Hence, future work is necessary to improve the absorptive capacity of this gel for recovering acids and alcohols. The endeavors could include structural modification of the gel to attain better selectivity of the alcohols. Due to the easy portability and small amounts of gel required, it can be used for in-situ recovery of the alcohols. Due to the property of recyclability (Chakraborty et al., 2019c), it can be reused for continuous operation, in-line with the reactor. Figure 7.3 represents future applications of waste valorization techniques which could be followed for maximizing the development of this approach to produce platform chemicals such as volatile fatty acids and alcohols.

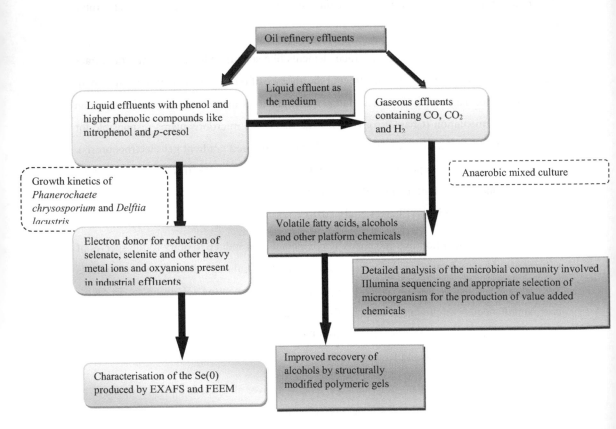

Figure 7.3 Schematic showing the future perspectives of this research

A bidirectional approach of valorizing the liquid and gaseous wastes of the petrochemical industry was demonstrated in this thesis. The aerobic biodegradation of wastewater containing phenol by two different co-culturing techniques using *Phanerochaete chrysosporium* and *Delftia lacustris* was developed. Simultaneously, the selenite present in wastewater was reduced to nano Se(0) which finds potential applications in the pharmaceutical industry. CO was used as the model pollutant, representing gaseous emissions of oil refinery and petrochemical industry, to produce acids and alcohols, including ethanol, butanol and hexanol.

The highest concentration of hexanol produced in this study was 1.46 g/L. In the absence of tungsten, an initially high pH of 6.2 followed by a pH decrease to 4.9 yielded 7.34 g/L of acetic acid as major acid at the end of the high pH stage. Subsequently, bioconversion of the acids at a lower pH of 4.9 yielded 1.85 g/L ethanol and 1.2 g/L butanol in the absence of W. A similar assay with two consecutive periods of pH, maintained at 6.2 first and then decreased to 4.9, yielded 6.6 g/L of acetic acid at high pH, 4 g/L of ethanol as well as 1.88 g/L butanol at the lower pH in the absence of selenium. To recover the desired biofuel, i.e. alcohols particularly ethanol, an eco-friendly, cost-effective co-polymeric gel was synthesized, which was found to be convenient in terms of storage and reusability.

References

Abubackar, H. N., Veiga, M. C., Kennes, C. Carbon monoxide fermentation to ethanol by *Clostridium autoethanogenum* in a bioreactor with no accumulation of acetic acid. Bioresour. Technol. 186 (2015): 122-127.

Azad, A.K., Rasul, M.G., Mofijur, M., Bhuiya, M.M.K., Mondal, S.K. Sattar, M.K. Energy and waste management for petroelum refining effluents: A case study in Bangladesh. Int. J. Automot. Mech. Eng. 11 (2015): 2170-2187.

Campbell, C.J. Petroleum and people. Population and Environment, 24 (2002): 193-207.

Chakraborty, S., Rene, E.R., Lens, P.N.L. Reduction of selenite to elemental Se (0) with simultaneous degradation of phenol by co-cultures of *Phanerochaete chrysosporium* and *Delftia lacustris*. J. Microbiol. 57 (2019a): 738-747

Chakraborty, S., Rene, E.R., Lens, P.N.L, Veiga, M.C., Kennes, C. Enrichment of a solventogenic anaerobic sludge converting carbon monoxide and syngas into acids and alcohols. Bioresour. Technol. 272 (2019b): 130-136.

Chakraborty S., Bera, R., Mandal A., Dey, A., Chakrabarty D, Rene E.R., Lens P.N.L. Adsorptive removal of alcohols from aqueous solutions by N-tertiary-butylacrylamide (NtBA) and acrylic acid co-polymer gel. Mater. Today. Commun. (2019c): (In press)

Elnaker, N. A., Elektorowicz, M., Naddeo, V., Hasan, S. W., Yousef, A. F. Assessment of microbial community structure and function in serially passaged wastewater electro-bioreactor sludge: An approach to enhance sludge settleability. Sci. Rep. 8 (2018): 7013-7023

Ellwood, M.J., Schneider, L., Potts, J., Batley, G.E., Floyd, J. Maher, W.A. Volatile selenium fluxes from selenium-contaminated sediments in an Australian coastal lake. Environ. Chem. 13 (2016): 68-75.

Fernández-Naveira, Á., Veiga, M.C., Kennes, C. Selective anaerobic fermentation of syngas into either C2-C6 organic acids or ethanol and higher alcohols. Bioresour. Technol. 280 (2019): 387-395.

Liu, K., Atiyeh, H.K., Stevenson, B.S., Tanner, R.S., Wilkins, M.R., Huhnke, R.L. Continuous syngas fermentation for the production of ethanol, n-propanol and n-butanol. Bioresour. Technol. 151 (2014): 69-77.

Liu, J., Lin, Q., Chai, X., Luo, Y., Guo, T. Enhanced phenolic compounds tolerance response of *Clostridium beijerinckii* NCIMB 8052 by inactivation of Cbei_3304. Microb, cell fact. 17 (2018): 35.

Mayfield, S., Wong, P.K. Forum Chemical engineering: Fuel for debate. Nature, 476 (2011): 402.

Mal, J., Nancharaiah, Y.V., Van Hullebusch, E.D., Lens P.N.L. Metal chalcogenide quantum dots: biotechnological synthesis and applications. Rsc Adv. 6 (2016): 41477-41495.

Psomopoulos, C.S., Bourka, A., Themelis, N.J. Waste-to-energy: A review of the status and benefits in USA. Waste Manage. 29 (2009): 1718-1724.

Salcedo, G.M., Kupski, L., Degang, L., Maree, L.C., Caldas, S.S., Primel, E.G. Determination of fifteen phenols in wastewater from petroleum refinery samples using a dispersive liquid—liquid microextraction and liquid chromatography with a photodiode array detector. Microchem. J. 146 (2019): 722-728.

Saxena, J., Tanner, R.S. Effect of trace metals on ethanol production from synthesis gas by the ethanologenic acetogen, *Clostridium ragsdaleii*. J. Indus. Microbial. Biotechnol. 38 (2011): 513-521.

Sadrimajd, P., Rene, E.R., Lens, P.N.L. Adsorptive recovery of alcohols from a model syngas fermentation broth. Fuel. 254 (2019): 115590-115597.

Sheng, G.P., Yu, H.Q. Characterization of extracellular polymeric substances of aerobic and anaerobic sludge using three-dimensional excitation and emission matrix fluorescence spectroscopy. Water Res. 40 (2006): 1233-1239.

Wadgaonkar, S.L., Nachariah, V.Y., Jacob, C., Esposito, G., Lens, P.N.L. Selenate reduction by *Delftia lacustris* under aerobic condition. J. Microbiol. (2019) (In Press).

Werkeneh, A.A., Rene, E.R., Lens, P.N.L. Simultaneous removal of selenite and phenol from wastewater in an upflow fungal pellet bioreactor. J. Chem. Technol. Biotechnol. 93 (2017): 1003-1011.

Young, L.Y., Rivera, M.D. Methanogenic degradation of four phenolic compounds. Water Res. 19 (1985): 1325-1332.

Netherlands Research School for the
Socio-Economic and Natural Sciences of the Environment

DIPLOMA

for specialised PhD training

The Netherlands research school for the
Socio-Economic and Natural Sciences of the Environment
(SENSE) declares that

Samayita Chakraborty

born on 1 January 1990 in Kolkata, India

has successfully fulfilled all requirements of the
educational PhD programme of SENSE.

Paris, 12 December 2019

The Chairman of the SENSE board

Prof. dr. Martin Wassen

the SENSE Director of Education

Dr. Ad van Dommelen

The SENSE Research School has been accredited by the Royal Netherlands Academy of Arts and Sciences (KNAW)

KONINKLIJKE NEDERLANDSE
AKADEMIE VAN WETENSCHAPPEN

The SENSE Research School declares that Samayita Chakraborty has successfully fulfilled all requirements of the educational PhD programme of SENSE with a work load of 61.1 EC, including the following activities:

SENSE PhD Courses

- o   Environmental research in context (2016)
- o   Research in context activity: 'Co-organizing ETeCOS3 and ABWET PhD Summer School (23-27 May 2016, Delft) and COST Training School (25-27 May 2016, Delft)'

Other PhD and Advanced MSc Courses

- o   The ABWET introductory course on anaerobic digestion, IHE Delft (2016)
- o   ETeCoS3 and ABWET Summer School on Contaminated Sediments - Characterization and Remediation, IHE Delft (2016)
- o   3rd Summer School on Biological Treatment of Solid Waste, University of Cassino (2017)
- o   Orientation to doctoral studies, Tampere University (2017)
- o   Research Ethics (2017) & Scientific writing (2018), Tampere University
- o   Entrepreneurship and Innovation (2017) & How to write a proposal (2017), University of Cassino
- o   Biological wastewater treatment: principles, modelling and design, TU Delft and IHE Delft (2018)
- o   Biogas technology for material flow management and energy production, Tampere University (2018) & Industrial organic chemistry (2018), Tampere University
- o   Summer School in advanced waste to energy biotechnologies: bio and circular economy approaches, Tampere University (2018)

External training at a foreign research institute

- o   Handling toxic gases like CO, UDC, La Coruña, (2017-2018)
- o   Microbial DGGE analysis, Tampere university of applied sciences (2018)

Management and Didactic Skills Training

- o   Supervising MSc student with thesis entitled 'Optimisation of alcohol adsorption on polymeric resins: kinetics and neural network modelling' (2017)

Oral Presentations

- o   *Bioconversion of CO and CO2 to biofuels and bioelectricity.* 1st International ABWET Conference : Waste-to-bioenergy : Applications in Urban areas. 19-20 January 2017, Paris, France
- o   *Production of C2-C6 alcohols from CO by anaerobic sludge,* G16-ABWET conference 6-7 December 2018, Naples, Italy

SENSE coordinator PhD education

Dr. ir. Peter Vermeulen

Printed and bound by CPI Group (UK) Ltd, Croydon, CR0 4YY

24/10/2024

01778295-0001